IMPACT EARTH
Asteroids, Comets and Meteoroids
The Growing Threat

IMPACT EARTH

Asteroids, Comets and Meteoroids
The Growing Threat

Austen Atkinson

First published in Great Britain in 1999 by
Virgin Publishing Ltd
Thames Wharf Studios
Rainville Road
London W6 9HT

A catalogue record for this book is available from the British Library.

ISBN 1 85227 789 0

Phototypeset by Intype London Ltd
Printed and bound in Great Britain by
CPD Wales

CONTENTS

List of illustrations   ix
Acknowledgements   xi
Foreword by Dr Brian Marsden   xiii
Prologue   xxi

Part I

1 The Nature of the Threat   3
An overview of the impact problem. JA1: the largest object to approach Earth since records began. A scientist's plea to John Major. British and American concerns. The prospect of an impact winter. Monica Grady's calculations of energy release from an impact. The Barringer crater in Arizona. The dinosaurs: what happened? Napier and Clube and the Alvarez team: extinction by impact. The Chicxulub mystery. Does the humble fern give us a clue? Tree-ring patterns and ice-core records provide a chronology. The Greenland Ice Sheet Project II (GISP2). Tunguska and the Beta-Taurid asteroids. Impact catastrophe: a VR simulation.

2 Engine of Evolution, Destroyer of History   27
Concerns of Shoemaker following Jupiter impact. Panspermia theory of life on Earth. Chiba's model for long-term patterns of evolution. Neanderthal Man side by side with Modern Man? Clube and Asher: cooling by cometary debris. Threat from the Beta-Taurid meteor stream. Tree-ring data as evidence of catastrophe. Cassiodorus's 'deeply curious' mystery in a sixth-century sky. Was Phaethon's fiery chariot an impactor? Atlantis and other lost civilisations. 'Little Ice Age'. Migrations due to atmospheric dust? Have heavenly bodies steered human history?

3 Ages of Ice and Water   49
Earth's many ice ages: were impacts the cause? Clube's Beta-Taurid

hypothesis for last ice age. Milankovitch oscillations as Earth hurtles through space. Volcanic loading of the atmosphere and Mount Pinatubo. Global warming: more to come? The predators that are no more. The passing of Neanderthal Man. Floods of biblical proportions due to global temperature fluctuations. The Eltanin impact of 2.16 million years ago. Ovid's flood account and the Great Flood of Genesis. The 'Out-of-Africa' theory. How would we cope with an ice age?

4 Tomorrow Is Yesterday   70

History is changed by bodies from space. Space station Alpha: hope for the future? Montezuma's Aztec empire and the portent of doom. 1066 and all that: Harold's omen in the sky. Other omens that shaped history. Hale-Bopp and the Heaven's Gate cult. Tunguska: the day the Earth was hit. Halley's Comet: an eyewitness account. Terror in the Brazilian jungle. The Jupiter impact and calls for further research into deathrocks. EU declaration on near-Earth objects: a breakthrough? Children die in Bogotá impact. A meteor crashes into Greenland's ice cap. Are our governments inactive or inept?

5 Lies, Damned Lies and Conspiracy   90

Are governments hiding information? Star Wars: could SDI become a sword-turned-ploughshare? The 1979 Agreement Covering the Activities of States on the Moon and Other Celestial Bodies. Government obfuscation on BSE and deaths from radiation. MOD's bland response to a concerned scientist. Why the conspiracy of silence? Did SDI detect Saddam's Scuds? And what is Britain's connection? HAARP: towards a greater understanding of aurorae. Tesla's experiment to harness the ionosphere. The NEAT programme to detect potential killer rocks. Asteroids as a strategic weapon, made possible by knowledge gap. Natural Environment Research Council's poor response to concerns. The dangers of governments' concern for status quo. Climate chaos. The soldiers and the scientists: cooperation at last.

6 Shielding the Cradle   113

Nostradamus's 'King of Terror' prediction for 1999. What can we do to identify and deflect the deathrock? The probe to intercept Halley's Comet. The Rosetta mission and deflection technologies.

A nuclear solution and possible resistance from the international community. Each nation should have a detection system. Deep Space 1 and its ion engine. NASA's NEAR programme for asteroid rendezvous. SHIELD: an Earth-protection system. A proposal for protecting planet Earth. We need to take positive action now.

7 **Countdown to Collision**   134
A grim countdown to catastrophe. Demystifying space and making it 'ours'. Jim Benson and NEAP – a commercial journey into space. The corporate road to deflection technology. Should Europe go the Benson route? Luna-Sat: a microsatellite developed by a British firm. A shift in perception is emerging. XF11 and fears for an impact with Earth. The media reaction to XF11. Will XF11 still pose a problem? The effect on the public: greater awareness of the threat. The Internet: a conduit for information. US initiatives to boost understanding. What about the southern hemisphere? A role for amateurs. Spaceguard. The young will win the argument. *Clementine II*. A public–private initiative to SHIELD the world? We must save our Eden.

Part II

I: Time to Die: a 'factional' impact scenario   159

II: Aeons of Fire and Ice: 'factional' consequences of impact   194

Afterword by Major Jonathan Tate   217

Appendix   223

Glossary   237

Index   261

LIST OF ILLUSTRATIONS

The Near Earth Asteroid Rendezvous (NEAR) spacecraft (© NASA)

A NASA artist's impression of an 800-kilometre asteroid striking the Earth (© NASA)

3-part computer simulation of a Shoemaker-Levy 9 type comet impacting in Earth's oceans (© Sandia National Laboratories)

Computer graphics image of the 65-million-year-old Chicxulub crater (© VL Sharpton, LPI)

Cenote Ring imposed on a mosaic of Landsat images, revealing the location of the Chicxulub crater (© NASA)

Aerial view of a cenote (© NASA)

SST sea-temperature satellite picture 11 August 1996 (© National Oceanographic and Air Administration)

SST sea-temperature satellite picture 11 August 1997 (© National Oceanographic and Air Administration)

Stone tablet c. 87 BC – one of the oldest human records of Halley's Comet (© British Museum)

Roman commemorative coin (© American Numismatics Society, New York)

Sketch of the solar system by Pedegache (© British Library)

Predictive simulation of the impact fireball created by the Shoemaker-Levy 9 impact with Jupiter (© Sandia National Laboratories)

Comet Shoemaker-Levy 9 impacting with Jupiter, July 1994 (© NASA)

Gene Shoemaker (© Lowell Observatory)

Comet Shoemaker-Levy 9, 30 March 1993 (© NASA)

Comet Hale-Bopp photographed over Avebury (© Linda Whitnell, Little Gable, 33a Wycombe Road, Holmer Green, High Wycombe HP15 6RX)

ACKNOWLEDGEMENTS

The journey from inspiration to finished manuscript has been a long one and I have encountered many extraordinary individuals along the way who have contributed to the whole. Donna, my wife, is exceptionally long-suffering. Despite countless acts of patience above and beyond the call of duty over the years, she continues to enthuse about my work and is my best friend and most ardent supporter. Her keenly analytical mind, that of a zealous biochemist, has long been an object of my fascination and of tremendous help in a sticky scientific moment.

My parents and grandmother are all quite mad but determined to cling to their own picture of reality, and in so doing have given me a valuable and unusual perspective on life. My mother is a human dynamo and I owe my approach to my working life to her; while my father has been my fiercest critic and greatest fan throughout.

My friends are nearly as long-suffering as my poor family; I would like to offer thanks to them for tolerating my darker moods – especially Chris James and Paul Waddington.

Few connected with the bizarre fantasy world known as publishing understand the true importance of the breed of talented liars and magicians generally known as literary agents. My own particular literary agent is Simon Trewin of Sheil Land Associates Ltd. Simon is one of Britain's best literary magicians and is in fact a good man and a great friend. His words of wisdom have been of tremendous value during the preparation of this book, while the team at Sheil Land itself is second to none.

I have met, interviewed and learnt from some of the greatest minds on Earth, in their fields, over the last couple of years. While it would be impossible to thank them all individually, I would like to thank a core few in particular: Jasper Wall (late of the Royal Greenwich Observatory), Jonathan Tate (Spaceguard UK), Monica Grady (Natural History Museum), Jim Benson (SpaceDev), Robert Gold (Johns Hopkins University), Benny Piser (Liverpool John Moores

University), Ray Turner (Rutherford Appleton Laboratory) and Brian Marsden (Minor Planet Center). These amazing individuals, and countless others besides, have given freely of their time and allowed me to glimpse the enormity of their knowledge while helping me to develop my own perspective on the impact threat.

There are many colleagues in the journalistic and television professions whom I would like to thank, but they are too numerous to mention individually; suffice it to say that they know who they are, great men and women all. One word of special thanks goes to Sean Hargrave of the *Sunday Times* for his unswerving moral support.

Finally, I want to express my gratitude to the team at Virgin. Led by Rob Shreeve, Virgin Publishing is rapidly becoming one of the most innovative and daring publishers in the English language. Ray Mudie (Sales Director) has demonstrated remarkable perspicacity and has given this project his unswerving support. Most of all I want to thank the editorial team led by Rod Green, and in particular my editor, Anna Cherrett.

Anna is another one of those extraordinary individuals whom I mentioned earlier. She is a literary alchemist capable of coaxing a great performance from even the most war-weary and insufferable of authors (I must rank near the top of that list, unfortunately). Her sense of humour and long-distance moral support has been of inestimable importance, as have her incisive comments about the work itself. I am greatly in her debt.

If this book helps to raise the profile of the impact threat, to push it on to the political agenda, then the wonderful folk that have helped me draw my conclusions must accept some of the credit. Whatever the future holds, I am most grateful to you all.

FOREWORD

Several books on the danger to the Earth from impacts by asteroids and comets have appeared during the past few years. Nevertheless, the subject is vibrant with continuing developments in this surprisingly interdisciplinary field. The choice of the material in *Impact Earth* reflects this activity.

From a scientific viewpoint alone, the study of the impact hazard involves – among other disciplines – astronomy, physics, chemistry, geology, climatology, oceanology, ecology, biology, anthropology, archaeology, palaeontology and dendrochronology. The science tells us what is out there, what may be headed our way, what has impacted our planet in the past, and what the effect of past impacts was on the Earth, its atmosphere, its oceans and its life forms. Although events such as Chicxulub and Tunguska are nowadays widely known, the Earth is in fact peppered with impact craters, and some scientists have attempted to equate phenomena such as the Dark Ages and little ice ages with small impacts that were not directly observed.

The impact hazard is also important, of course, in terms of sociology, economics, politics and defence. The reaction of humankind to the possibility of an impact event is a dominant theme in Austen Atkinson's book – and rightly so. Sooner or later, there will be either an impact that will, or a demonstrated threat of an impact that would, drastically change the lifestyle of a substantial fraction of the Earth's population. Sixty-five million years ago, the dinosaurs were host to the ten-kilometre Chicxulub impactor, and the entire species suffered the consequences. In 1908, the fifty-metre impactor that exploded so violently over Tunguska, Siberia, was negligible in comparison but surely a reminder that danger from impacts is not just a thing of the distant past. We, too, could go the way of the dinosaurs. On the other hand, science could provide us with advance knowledge of the danger. How far in advance might that advance knowledge be? With only a matter of days, it might in fact be better that we not know, for our preparations would be largely futile. But, with the revelation of a specific event decades hence, the human race will be faced with a unique opportunity to preserve itself.

Are we up to the task? With a clear and identified threat, sufficiently far in the future, the military might of the world could be directed towards its removal. Whether this would succeed is at least as much a matter of politics and economics as it is of military technology. Politics and economics also play a strong role in whether and when any threat is identified. The necessary astronomical technology exists and is being used at some low level. Although the concern should be international, only a handful of countries are putting resources into this activity, and, in the countries that do participate, the emphasis is much more on merely counting the asteroids and comets discovered, rather than on examining whether one of the beans so counted will actually sprout disaster.

In March 1998 I was at the centre of the 'debacle' involving the prediction that an asteroid might pass very close indeed to the Earth in October 2028. The premise was innocent enough, and, at least initially, the press mentioned the story in a quite responsible manner. Why did the story even arise? The asteroid, called 1997 XF11, had been discovered three months earlier and was entered as No. 104 on the list of PHAs, or 'potentially hazardous asteroids'. This is a subset of the thousand or more vaguely defined NEOs, or 'near-Earth objects', the great majority of which can be no conceivable threat to the Earth during the next million years or so. Most of the PHAs give no cause for worry during the foreseeable future either, but they comprise a more manageable list for further investigation. Inclusion on the PHA list means only that the asteroid's orbit currently comes to within a few million kilometres of the Earth's and that the object is likely to be large enough for an impact to have a global effect. The 104 entries had accumulated since the discovery 65 years earlier of 1932 HA, now known as (1862) Apollo. With an estimated size of perhaps two kilometres and a particularly small orbit-to-orbit distance, 1997 XF11 should have been of greater than usual interest, but 1997 had already produced ten other PHAs, and rather little attention was paid to it. No attempt was made to determine its size or surface composition, for example.

Some routine follow-up measurements of the position of 1997 XF11 were made, but by early February 1998 most observers had given up on it, and the observations made at the McDonald Observatory in Texas on 3 and 4 March were the first in four weeks.

Although earlier computations had indicated a particularly close approach to the Earth in 2028, it was the inclusion of the data from 3 and 4 March that hinted at a miss distance of only tens of thousands of kilometres. Anybody who made the computation would have obtained this result, as, indeed, all those astronomers who later made the computation did. Sure, there was some uncertainty associated with the actual miss distance, but the tests that I made strongly suggested that the object would come closer than the moon. It became therefore a matter of trying to make up for lost time: ask observers to get more follow-up data (including data that could tell us the object's size), as well as search for possible images of 1997 XF11 on archival photographs taken in earlier years.

Although our message was to the international community of astronomers (professional and amateur), there was really no way it could be kept entirely secret. Indeed, it should not be so. As it is, there are too many accusations of 'cover-up' at the first intimation of danger. Whether there was really any danger from the asteroid was actually rather irrelevant – and the investigation of this was certainly of lower priority than seeking further observational data. If there had been any danger, it was the danger of apathy, and that had clearly now passed. This was an excellent illustration of setting science into action to obtain more data, thereby to solve the problem, one way or another.

Unfortunately, before further observational data could be obtained, other scientists made pronouncements to the effect that, on the basis of the existing data, the Earth was not in fact in danger of an impact. As far as the 2028 encounter was concerned, this was almost certainly true, but, by actually announcing this publicly, within hours of our issuing a request for further observations, and utilising numbers that were actually inconsistent, these scientists greatly confused the issue. Should I have examined the 2028 circumstances in detail before making my request for further observations? Possibly – but the object was getting more difficult to observe, and, if there was to be any hope of getting data on the object's size, time was of the essence.

But was the danger only to be in 2028? The Shoemaker-Levy comet passed close enough to Jupiter in 1992 to be tidally ripped apart. But only two years later it passed closer still – and all the fragments collided with the planet! To examine whether 1997 XF11 could be a danger to the Earth during the years and decades after 2028 would obviously take more time. The possibility of a subsequent impact does not even seem to have occurred to those who pronounced, 'That's zero probability of impact, folks!' On the contrary, they took the view that I should simply have consulted with them, with the assumption that we should achieve 'consensus' within, say, two days, on whether the Earth was or was not in danger. I fear this view is rather too simplistic. And what, pray, would happen if the consensus turned out to be that there was indeed 'a clear and present danger'?

As it happened, there *was* a danger, based on the observational data available at the time, that 1997 XF11 could collide with the Earth in the late 2030s or early 2040s. I began to appreciate this possibility a few days after the mid-March fiasco, although, for various reasons, more than two *months* (not two days!) passed before I was able to undertake the necessary calculations. The point is that the minimum distance between the orbits of 1997 XF11 and the Earth is decreasing at a moderately steady rate of some 4,000 kilometres per year. The minimum possible miss distance may well have been 30,000 kilometres in 2028, but what about a decade or so later?

Currently, 1997 XF11 takes about 1.73 years to orbit the sun. It was due to come close to the Earth in 2028 – conceivably as close as 30,000 kilometres, very likely closer than the moon, perhaps somewhat further away than that. The uncertainty in the miss distance translates rather directly into one in the object's revolution period after 2028. I ascertained that the subsequent period could have been as short as 1.53 years or as long as 1.99 years. So here we had the intriguing situation that, a decade or so after 2028, the orbit of 1997 XF11 would actually intersect the orbit of the Earth (the points of minimum separation spending perhaps three years within a distance equal to the Earth's radius), yet we had no idea where in its orbit the object would be!

On appreciating this, I knew that there had to exist some orbital trajectories, entirely consistent with the three-month arc of observations, that would indeed result in a collision between 1997 XF11 and the Earth. All I had to do was adjust the object's current 1.73-year period, at most by a matter of an hour or so, in order to make the post-2028 period resonant with the Earth to the point that this would minimise the miss distance during one of the years when the distance could be less than the radius of the Earth. In years, the post-2028 period therefore had to average a simple fraction in the designated range, like $\frac{5}{3}$, $\frac{7}{4}$, $\frac{8}{5}$, $\frac{9}{5}$, $\frac{11}{6}$, $\frac{11}{7}$ or $\frac{12}{7}$ – and, *voilà*, the numerator of the fraction would give the number of years from 2028 to a possible impact. I was able to ascertain that the $\frac{12}{7}$ case actually allowed a deep impact – on 26 October 2040. The $\frac{9}{5}$ case gave a possible grazing impact on 27 October 2037.

Most of the other fractions above, as well as others with numerators up to 17 and more, easily yielded miss distances of only two or three Earth radii, and I should not have wanted to say that a collision could not have occurred. In at least some of these years, the probability of impact by 1997 XF11 would be approaching 1 in 100,000, which is two orders of magnitude greater than the accepted annual impact rate for unknown two-kilometre asteroids.

Fortunately, of course, we don't have to worry about an impact by 1997 XF11, because my announcement quickly (almost too quickly!) had the desired effect, and images of the asteroid were recognised and measured on photographs taken in 1990. These observations showed unequivocally that the miss distance in 2028 will in fact be almost a million kilometres, and there is no possibility of any other approach within some 3 million kilometres during the whole of the next century. By then, of course, the minimum distance between the orbits of 1997 XF11 and the Earth will have increased again to some 200,000 kilometres, and one can confidently remove this object as any threat for a very long time in the future.

If the actual 2028 miss distance of 1997 XF11 is 1 million kilometres, some two and a half times the distance of the moon, why was there such confidence, in the absence of the 1990 data, that it would come *closer* than the moon? A careful analysis by the Finnish astronomer Karri Muinonen (which also took several months to complete) concluded that the probability of a distance less than that of the moon was as high as 72 per cent. Furthermore, the probability that it would come closer than it actually does was a startling 99 per cent! Sometimes one can be a victim of unfavourable statistics. This is also why we need to consider that a two-kilometre asteroid could hit the Earth during the next hundred years or so, even though such impacts are typically spaced by intervals of perhaps a hundred thousand years.

During 1998 there were four times as many PHAs discovered than in any earlier year, resulting in a 50-per-cent increase in the total number of known PHAs. This dramatic augmentation, mainly due to the 'LINEAR' programme funded by the US Air Force, is a wonderful affirmation of the importance of the Earth-impact threat. The US National Aeronautics and Space Administration (NASA) has also pledged a significant increase in funding for NEO searches. But, as I have noted, merely to count the PHAs and NEOs found is not sufficient. With only 10 per cent of the kilometre-sized dangers already detected, there is obviously a need for more search work, particularly from the southern hemisphere, which is completely devoid of NEO search activity.

But we also need to examine carefully what we find. The principal message of 1997 XF11 is that we must not be complacent and assume that what we think we know cannot possibly hurt us. To 'know' a PHA we need to be able to link together observations made in different years. Sometimes we shall be lucky and find images on old photographs – but only if some of the funding goes into this enterprise. More generally, second-year follow-up has to be made when the object is far away and faint, and this requires the availability of telescopes larger than those involved in the basic searches. Large telescopes are also frequently needed for physical data – although for 1997 XF11 we might as well now wait until 2002, when it comes to a distance of some 10 million kilometres and will be accessible to small telescopes.

The amount of time Muinonen and I spent on examining the nontrivial aspects of the 1997 XF11 situation, long after the 1990 observations solved the problem, shows that meaningful conclusions can sometimes be drawn about the danger from specific PHAs, even while one is awaiting further observational data. Actually, the 1997 XF11 analysis is incomplete, and we still lack a rigorous estimate of the impact probability during the next half-century – prior to the discovery of the 1990 data, that is. In view of the chaotic nature of the object's orbit, this is a problem that is almost intractable. There is no other asteroid observed over more than a fraction of a day for which collision orbits over a matter of decades have been demonstrated.

An interesting recent study by the Polish astronomer Grzegorz Sitarski of (4179) Toutatis, a well-observed asteroid somewhat larger than 1997 XF11 that will pass less than 2 million kilometres from the Earth in 2004, and of which there exist images from as long ago as 1934, shows that the chaotic motion is such that an Earth impact a few centuries from now cannot be excluded. This study considers the asteroid's motion also to be influenced by the nongravitational forces one associates with cometary outgassing, and it is not in fact clear that they apply in this case. Nevertheless, there are active comets with chaotic orbits for which they do apply, and it is necessary to consider that some of these bodies are threats.

Despite the problem of orbital chaos, there is some reassurance in the fact that most of the impact danger necessarily comes from objects in short-period orbits that do not take them particularly far from the sun. The observational aspect of the problem is therefore largely tractable. To some extent, it would be possible to cull from the list of PHAs (and correspondingly for comets) a very select group of 'EDOs', or 'extremely dangerous objects', that clearly bears careful watching. The exception to all of this comes, of course, from the long-period comets that would have to be discovered on the same pass that yields the collision. With current technology, it is unlikely that an Earth-impacting long-period comet could be discovered more than a year or two ahead, and the warning time would probably be, at most, weeks.

There are differences of opinion as to the actual fraction of the danger posed by the long-period comets. Much of the difference seems to arise from considering the relative impact energy of the classes of object. Since the whole aim is to avoid an impact, this measure somehow seems inappropriate. Over a time interval of a few centuries, an object with a short revolution period may have numerous opportunities to get us. Object for object, the fraction of the danger due to long-period comets can scarcely be larger than perhaps 2 per cent.

As an astronomer, I tend to regard the impact problem as one that is largely astronomical. Ultimately, of course, it may become a matter of defence. Let me repeat: are we up to the task? Now you should read Austen Atkinson's book and draw your own conclusions.

Dr Brian G Marsden
Director
Planetary Sciences Division
Harvard-Smithsonian Center for Astrophysics

December 1998

I saw a star fall from heaven unto the earth . . . and there arose a smoke out of the pit, as the smoke of a great furnace; and the sun and the air were darkened by reason of the smoke of the pit.

— Revelation 9:1–2

PROLOGUE

This is a book about the imminent destruction of all life on Earth. Life is a property that is little understood: the defining spark is a mystery to late-twentieth-century scientists, its origin still a matter for philosophers and clergy.

Perhaps there is no reason for alarm if life on planet Earth, currently amounting to countless billions of arthropods, birds, fish and mammals, is deemed to be of no importance in the scheme of the universe. Or perhaps the importance of this global mass death may depend upon the sort of life we are considering. By far the most populous form of living creature is that of the arthropod, especially those belonging to the *Insecta* class. At least 1 million species have been recognised; the *Insecta* class extends to 6,000 million million – approximately 1 million insects for every human being on Earth.[1] Their form has existed for at least 400 million years, the earliest fossil remains dating from the Devonian period (circa 408–350 million years ago). An observer not familiar with Earth might, therefore, come to the conclusion that *Insecta* is, unquestionably, the ruler of the planet. An understandable conclusion and possibly one shared by the insects, though until we develop technology that will allow us to ask them their opinion we are unlikely to know their views for certain. Would a mass outcry of anger and grief ensue if all insects were to die, subjected to a mass extinction thanks to a catastrophe? Probably not. For, despite their vital place in the food chain and active role in the health of the biosphere, insects are not very friendly to non-insects, nor are they very cuddly. So perhaps no one would mourn the passing of this remarkable and resilient class of organisms.

Should we worry about the mass extinction of other species? A relatively new group of creatures, with less than 100,000 years of

[1] McEwan, P (Dr). *Science Fiction Becomes Science Fact as UK Homes Become Overrun With Insecticide-Resistant Bugs*, Consolidated Communications/Insect Investigations Ltd, 1998.

ancestry, in its current form, has emerged as would-be ruler, being even more dynamic than the insects. True, its social system is less cohesive and ordered than that of the insect, its reproductive system is slow and complex and its ability to withstand pestilence is negligible; yet it has risen to rival the insect and indeed surpass it in several key ways. With a worldwide population that amounts to only 6,500 million, it has developed a societal infrastructure that, although riddled with dissension, violence and war, a state unknown to insects, can literally change the surface of the planet itself. Techniques for destroying tracts of land, infecting other areas with its own form of pestilence to the point of poisoning whole rivers, or even stripping a forest of its trees, have all been developed in less than 100,000 years. A remarkable feat.

It is this ability to control, alter and exploit the Earth on a scale never before attempted by any other animal or plant life that makes this species the prime candidate for the title of ruler of planet Earth. That species is a land-based creature, part of the class known as *Mammalia*: it is a creature that nurtures its young via mammary glands, is warm-blooded, has bodily hair, seven vertebrae, a jaw formed by just one bone and its red blood cells function without a nucleus.[2] The particular class of mammal in question is *Homo sapiens sapiens*, a new descendant from the ape subset of mammals.

Humankind, as it calls itself, has come to perceive its place in the universe in a number of different ways in recent history – progressing from a geocentric view (one that believed that the Earth was the centre of the solar system) to a heliocentric one (which correctly acknowledged the sun as the centre of the solar system). Apparently emerging in the Bronze Age,[3] when *Homo sapiens sapiens* first worked metal ores extracted from the land, circa 3500–500 BC, the concept of 'kingship' and gods has played an important part in humankind's comprehension of creation: the development of life on Earth and its demise. Geocentrists – early scientists attempting to correlate the world they saw around them with the established 'facts' about creation as presented in the Bible – believed that all celestial bodies must orbit the planet Earth. They saw Earth as the centre of

[2] Ed. Crystal, David, *Cambridge Encyclopaedia*, Cambridge University Press, 1994.
[3] According to the theories of Gunnar Heinsohn, University of Bremen.

the universe and therefore the most important part in what they perceived as God's creation.

An intellectual revolution, begun by the Polish rational thinker Dr Nicolaus Copernicus (1473–1543) around 1512, switched human-kind's concept of its place in the universe to a heliocentric one.[4] Copernicus realised that the dogma of previous centuries, rationalised by Ptolemy (c.90–168) into a cosmos of perfectly circular orbits around the centre of all things (Earth), was clearly wrong. His theory, first published in 1543, changed perception, albeit at the expense of Copernicus's reputation and soul: the Roman Catholic Church damned the book as heretical, issuing a holy ordinance against it in 1616, not accepting the basic tenets of the theory until 1835, 300 years after it was published. Theoreticians and philosophers lived in true fear at this time: they were far from free to propound their concepts, and occupied a position in society that can be likened today only to that of a political radical.

Whereas today the grinding engine of democratic process and the inherent time-lag necessitated by the long educational period of human generations act as a limit to the dissemination of knowledge, in the Middle Ages the Inquisition was the establishment's main tool of knowledge suppression. Operating in various forms for more than 700 years, the Inquisition was a judicial body established under Roman Catholic papal instruction, with the role of eradicating heresy, witchcraft and other deviations from strict contemporary doctrine. Following the authorisation of Innocent IV, torture was employed as a means of securing confessions to heresy – often from innocent individuals. The Inquisition was responsible for enforcing the suppression of original scientific thought, not least Copernican views and those of his followers (Galileo *et al*).

The Inquisition was renamed the Holy Office in 1908 following Pius X's decision to modernise the institution. Pope Paul VI instigated further reform, and in 1965 rechristened the infamous body the Congregation for the Doctrine of the Faith. Its role was softened to one of ensuring Catholic adherence to strict doctrine. In its time the Inquisition has been most effective in its pursuit of ignorance and the suppression of science, even in the twentieth century, irrespective of the sheep's clothing it has worn since the 1890s.

[4] Gribbin, J, *Companion to the Cosmos*, Weidenfeld & Nicholson, 1996, p. 88.

The blame for the horrors of the Inquisition must lie with St Augustine of Hippo (354–430) and Pope Gregory IX (1170–1241). Augustine liberally reinterpreted the gospel of St Luke (14:23) as suggesting that the use of force should be countenanced against all heretics. The passage reads, 'And the Lord said unto the servant, Go out into the highways and hedges and compel them to come in, that my house may be filled.'

Pope Gregory IX's act of formally establishing the Inquisition as an organisation and deploying fanatics as anti-heretical thought police, while justifying his act as fulfilling his role as God's representative on Earth, must rank him as one of society's greatest ever masters of 'spin'.

Galileo Galilei (1564–1642), an exceptional Italian scholar, mathematician and astronomer, was sentenced to indefinite imprisonment by the Inquisition in 1633, charged with expounding theories about planetary motion. Despite recanting all of his hypotheses at his trial, he was denounced as a heretic. It is evident that he believed his theories about planetary motion to be correct and his trial to be an exercise in suppressing truth, as, following a public renouncement of his views at his trial, he is reputed to have uttered, ' . . . and yet it [the Earth] moves!' (His prison sentence was later commuted by Pope Urban VIII to permission to reside at Siena.) Theoreticians literally risked their very lives for scientific truth, a point that we, as we move into the third millennium, must never forget.

The German astronomer Johannes Kepler (1571–1630) was the first true supporter of Copernicanism. A true geometrician, seeing mathematical splendour in the order of the cosmos, Kepler grasped the mathematical truth in Copernicus's theory. Working at the University of Tübingen and later (1601) as the court mathematician of Emperor Rudolph II (a position he gladly took to escape the potential hazards of the Inquisition), he proposed the substitution of ellipses (1609) for the circular planetary orbits suggested by Copernicus. Kepler thereby explained the apparently erratic movements of the planets. He also created three laws of planetary motion which helped further rationalise the heavens.[5]

Kepler's modified Copernican model was given mathematical

---

[5] Kepler, J, *Epitome Astronomiae Copernicanae*, c. 1620.

proof, in the seventeenth century, by Sir Isaac Newton's (1642–1727) application of the inverse square law to the Copernican question.[6]

Newton, undoubtedly one of the second millennium's greatest minds, theorised that the number of particles agglomerated in one place is directly proportional to the object's size of gravitational force: the larger the agglomeration of matter, the higher the gravity. Further, he theorised that the influence of a body's gravity extends over a very great distance, diminished to negligible levels only at a distance determined by an inverse square law. The law, commonly used in mathematics, dictates that the influence of a body, in this case gravity (g), which is assumed to have a spheroidal area of influence, varies in an inverse manner to the square of the distance (d) from that body. So, just as gravity can be said to increase as matter is added to an object such as a sun, so its sphere of influence can be said to diminish uniformly the further away one travels from it; the formula can be written simply, thus: $g = 1/d^2$. Newton's application of this fundamental mathematical principle to the problem of planetary orbits, by then observed using optical telescopes, further revolutionised *Homo sapiens sapiens'* view of the universe, resulting in a spectacular movement to rationalistic determinism that persists today.

Consequently, most of the *Homo sapiens sapiens* population, in the western hemisphere at least, has come to accept that the Earth itself has a rotational period of 23 hours, 56 minutes and 4 seconds, while completing an orbit of the centre of its solar system, an unremarkable sun, every 365.26 days. The Earth fits in place within a network of eight other major planets, each in turn orbiting about the sun. This solar system sits in a small part of a galaxy, known by *Homo sapiens sapiens* as the Milky Way, in a vast – probably expanding – universe of galaxies, each with approximately 100,000 million suns. *Homo sapiens sapiens* is, then, quite small by universal standards, its ancestral longevity barely a blip on the Earth's own geological time scale; and the Earth itself, thought to be approximately 4,600 million years old, is nought but a part of the record of universal time which extends back over approximately 20,000 million years.

Attempts to correlate biblical creationist theory, endorsed by the Inquisition, with geological fact led, in time, to the establishment of

[6] Newton, I, *Philosophiae Naturalis Principia Mathematica*, 1687.

a scientific doctrine known as catastrophism. The Frenchman Baron Georges Cuvier (1769–1832) could legitimately be described as the progenitor of palaeontology and comparative anatomy. Mindful of and committed to the Bible as the source of knowledge regarding the origin of Earth, he believed in the conclusion of Bishop Ussher (1581–1656) – Professor of Divinity at Dublin – that the Earth was created by God in 4004 BC. Ussher's life's work, *Annales Veteris et Novi Testamenti* (*Annals of the Old and New Testament*), established the date of creation – as Ussher saw it – and gave a full chronology of Scripture. Consequently his discoveries of fossil remains in the Paris basin – which clearly revealed breaks in the continuity of life on Earth – led him to endorse the long-established Judaeo-Christian belief that the world was created in its contemporary form only after facing a series of catastrophic episodes.

Although, in many ways, Copernican-Keplerian-Newtonian theory would endorse catastrophism, a form of rationalistic gradualism emerged in the seventeenth century which led to the formation of a different, secular, form of dogma in the nineteenth century: uniformitarianism. James Hutton (1726–97[7]), a noted Scottish scholar, established the basic principles of geology, especially the enormity of geological time and mechanisms of land formation such as sedimentary layers left by rivers developing into rock strata. He published his theories in 1795.[8] A contemporary Englishman, Sir Charles Lyell (1797–1875), embraced Hutton's concepts and coined the word 'uniformitarianism'. It is largely thanks to the efforts of these two men that the current sciences of palaeontology and geology exist as we know them today.

The product of understandably reactionary minds, uniformitarianism soon developed as the leading scientific doctrine, fostering Darwinian gradualism and suggesting that no catastrophe was involved with shaping the modern world – or at least not since the very beginning of Earth's history, which was vastly older than recorded history, eventually dated at 4,600 million years ago. Life on Earth had evolved via a process of survival of the fittest, encompassing slow adaptation to a very stable, barely changing

[7] Mutch, TA, James Hutton biography, Grollier Multimedia Encyclopaedia Version 10.0.0, 1998.
[8] Hutton, J, *Theory of the Earth*, 1795.

environment. Uniformitarianism is still, today, the dominant scientific rationale of the world around us.

Again I ask, is life on Earth important? Despite the acceptance and eventual modification of the Copernican theory and the acknowledgement that Earth is, comparatively, an infinitesimal body in an ocean of space so vast that no member of *Homo sapiens sapiens* can truly comprehend it – just one planet in orbit around a sun in a galaxy of several hundred billion stars – human life does indeed seem important. Why? Because 'intelligent' life on Earth may be unique. That may read like a testament to human arrogance, but, assuming that not one of the planets suspected to orbit the hundreds of billions of stars in this galaxy alone harbours any form of recognisable life, then Earth is special.

Homo sapiens sapiens may be the first creature to understand this concept in the history of the universe. The great *Insecta* has almost certainly never speculated on the importance of preserving itself or the human race, or any other form of life for that matter. Despite this lack of enthusiasm for ecology and welfare within the insect society, many human organisations dedicated to the preservation of animal and human life have emerged: the World Wide Fund for Nature, the Royal Society for the Protection of Birds, Oxfam, the Red Cross and Greenpeace, to name but a few. It would seem that these organisations, and the people who finance them via charitable subscription, are keen to preserve life in all its diversity. Life matters to them.

Churchyards – in truth necropoleis, yards of decomposed human corpses – are seen as places where extinguished life can be honoured: long-past moments of human happiness, achievement and endeavour are remembered. What are marked burial sites, common to all races of *Homo sapiens sapiens*, but a vast sign that life is of value? Elaborate ceremonies have developed to herald a new life, a human birth, including praise unto 'God' or the 'gods', a request to the transcendental one(s) for a safe and happy life and blessings by God's 'agents', priests or shamans, on Earth. It would seem, even from this cursory glance at human society in all its diversity, that *Homo sapiens sapiens* deems life of its own kind to be of value. And yet it is the insects who are most likely to inherit the Earth.

Animal life, even most current plant life, holds only a short tenancy on the Earth. Most mathematicians, if presented with the facts, would agree, I feel sure, that, despite *Homo sapiens sapiens'* inordinate tool-using sophistication, the species is most likely to face extinction sooner rather than later, if not by its own hand, then by that of 'God'. You might see it as carelessness on God's part, but all is not well in the universe. There exist a prodigious number of rogues, some small – little more than irritants to the surface of planets – others very large and of such power that they might obliterate the surface of a world.

Homo sapiens sapiens knows of them, of course: his tool-using skills have allowed him to develop advanced technologies capable of penetrating the solar system and observing the heavens with telescopes which can exploit visible light, X-rays, gamma rays, infrared and even radio emissions. Humankind has observed these rogues, with an enquiring eye, for more than two millennia, yet only at the end of the twentieth century is this would-be dominant species beginning to perceive the nature of the threat posed by these celestial wild cards. These rogues amount to a death warrant, already drafted and signed, merely awaiting the date of execution. If God did create them, and has some incredible mathematical model to record their every movement, then perhaps he might know when humankind's extinction will occur.

Currently, back on Earth, inactivity and indifference to the threat on the part of *Homo sapiens sapiens* seems likely to shorten the odds of his extinction. The threat itself can be broken down into several levels: destruction of a small area or city, perhaps 2,000 square kilometres; loss of life on a continent and global environmental instability; or ultimately a mass-extinction event, defined by human scientists[9] as the destruction of more than 25 per cent of all life on Earth. The instruments of this threat are as diverse in nature as they are numerous. They can, however, for the ease of dissemination, be placed into three main categories: asteroid, meteoroid and comet.

Asteroids are rubbish. By that I mean waste, leftover raw materials from the creation of the solar system, probably. No one knows for certain, but it seems the most likely explanation. Most asteroids known to science sit in identifiable orbits, in a huge ring – generally referred to as the asteroid belt – between the orbits of Mars and Jupiter. Assuming an Astronomical Unit (AU), the measurement used to describe distance in the solar system, to be equivalent to the

[9] Morrison, D, Chapman, CR, and Slovic, S, *Hazards Due to Comets and Asteroids: The Impact Hazard*, University of Arizona Press, 1994, pp. 59–89.

distance of Earth from the sun – approximately 150 million kilo-metres – then the asteroid belt stretches from 1.7 to 4 Astronomical Units from the sun. In other words, the asteroid belt is so wide that it occupies 2.3 times more space than the distance between Earth and the sun.

Asteroids are dead, in geological terms, having no volcanic activity whatsoever. Cold, barren mountains of rock and rubble in space. Small on a planetary scale they may be, but some asteroids are enormous on a human scale: Ceres, the largest yet observed, has a diameter of about 930 kilometres. Put another way, traversing Ceres' diameter would be like walking from Stonehenge on Salisbury Plain, England, to the Culloden Battlefield, near Inverness, Scotland. Their composition has been determined via the collection of samples that have fallen to Earth and via spectroscopy – measuring the properties of light reflected from the mineral components on the surface of an object – of asteroids in space. Each mineral has either a known corresponding light-absorbing or -emitting property, knowl-edge gained by laboratory experiment, which can be measured by instrumentation mounted aboard a satellite or spacecraft, or from Earth-based instrument arrays. Light is a form of radiation, com-prising the electromagnetic spectrum, and not all parts of the spectrum are visible, so astronomers also use invisible wavelengths of light to measure the properties of space-based objects. Asteroids are composed of a variety of materials, the vast majority fitting into three broad categories: silicates (stony), iron silicates (stony iron) and nickel-iron alloys. Taenite and kamacite are two minerals rarely found on Earth, but they occur with great frequency in asteroids.

Meteoroids are thought to be Earth's most frequent form of extra-terrestrial visitor. Little different from asteroids, they are thought to originate from asteroid collisions, debris thrown off into Earth-crossing orbits by the impact. Meteors are common visitors to the night sky. Otherwise known as shooting stars or fireballs, they occur when a small meteoroid (a few tens of metres) hits the upper atmos-phere, travelling at speeds of 20–70 kilometres per second (km-1) and friction with the air results in a fireball. The stony meteorites are very difficult to find on Earth, because their composition is so similar to that of terrestrial rocks that their disguise could be said to be perfect. In fact, few stony meteoroidal bodies are dense enough to impact intact, so the stony iron and iron meteorites are the most commonly found examples.

Meteors can fall as part of a shower or storm, such as the Leonid shower, which produces a storm on a semiregular cycle, approximately once every 33 years. Indeed, it was the first recorded Leonid shower in 1833, so called because they appear to emanate from the constellation of Leo, that led directly to the scientific study of the meteorite phenomenon for the first time. In a highly advanced world dependent on orbital military, scientific, telecommunications, global-positioning and other, commercial, satellites, meteoroidal impact does pose a major threat. A minor impact with an artificial satellite can result in an electrical surge which can disable a craft's electronic systems, effectively rendering it a useless piece of expensive junk.

The single greatest threat to humankind may well be that of the rogue comet. Described in 1949 by the American scientist Fred L Whipple[10] (b. 1906) as dirty snowballs, the comet presents one of the most beautiful, awe-inspiring sights in the heavens. Invariably vast, these great balls of ammonia, dust, rock and methane fall into two categories: short-period and long-period comets. Short-period comets have an orbit that allows them to return to Earth's vicinity in less than 200 years. These comets are largely identified, their orbits known, though some may still remain undetected. Long-period comets, however, are far more dangerous. Often referred to as rogues, these comets have orbits that take them away from Earth for perhaps thousands of years. Their angle of incursion, or descent, towards Earth might be at any angle to the orbital plane of the planet (known as the ecliptic), making detection particularly difficult. Consequently, they may 'appear' in very close proximity to Earth, with only a few hours' warning – or, more frighteningly, no warning at all. They travel at an average speed of 38 kilometres per second,[11] and are invariably visible, when in close proximity to the sun, thanks to two huge tails – an ion and a dust tail – given off when solar winds (a flow of energised particles from the sun) heat the frozen cometary nucleus causing it to vent melted matter in large jets. The tails can stretch to 1 AU, the distance between Earth and the sun. Because the tails are generated by the solar winds, they always point away from the sun, until the comet's orbit carries it far enough away from the

---

[10] Whipple, FI, *The Mystery of Comets*, Cambridge University Press, 1986.
[11] Grady, M, interview, 1998.

sun's influence to allow its nucleus to stop outgassing and return to a quiescent state.

Comets are vast. They move incredibly quickly, and long-period comets are being discovered at the rate of one per month, suggesting to astronomers that there could be some 12,000 million comets orbiting in a vast mobile spherical mass – known as the Oort cloud, after its proposer, Jan Oort (1900–92) – some 50,000 AU from the sun, at the edge of our solar system. A smaller belt amounting to 1,000 million comets, first suggested by the cometary theorist Gerard Kuiper (1905–73) and known as the Kuiper Belt, may orbit the sun at a distance of 35 to 1,000 AU. This belt could be significantly destabilised by the gravitational pull of the giant gaseous planets such as Jupiter – which have enormous mass and therefore huge gravitational fields – and be thrown into the centre of the solar system. This theoretical process may account for the creation of all the comets that wander through the solar system, many of which visit near-Earth space on a regular basis. An example of one of these is Halley's Comet, which approaches Earth approximately every 76 years and was last seen in 1986.

If a one-kilometre rogue comet, small by any standard, were to hit one of Earth's oceans, it would generate a release of energy so massive that it would be akin to that generated by 300,000 million tons of TNT.[12] To place this threat in perspective, the Hiroshima nuclear bomb had an equivalent explosive capacity of just 12,000 tons of TNT. Uniformitarian doctrine has no place for this research and yet scientists around the world are engaged in studies into the potentially cataclysmic effects of cometary, asteroidal and meteoroidal impacts.

True uniformitarian doctrine is as absurd as traditional catastrophism: both models of the world around us are too simplistic – they creak and buckle under the weight of modern research and data like the knees of an obese dinosaur. The truth must lie in a synthesis of the two models, adapted and evolved to incorporate new evidence of frequent catastrophic episodes developed since the 1970s: gradualism with geological moments of extreme catastrophe and change. This form of neo-catastrophistic-uniformitariansim, which might be more happily called adaptive gradualism, is not some form of atavistic redress, but a pragmatic acceptance of reality; just as Copernican

[12] According to Sandia National Laboratories' teraflop simulations.

theory was a pragmatic rationalisation of reality as Nicolaus Copernicus perceived it, so too is this new synthesis.

Our Earth has probably faced global decimation on no fewer than five occasions in the past: 11, 35, 65, 92 and 210 million years ago. Probably tens of thousands of major perturbations have rocked various biomes (an ecological community stretching over a large area) since *Homo sapiens sapiens* first walked the Earth; indeed his very existence may be as the result of an evolutionary surge needed to survive a catastrophic impact. Even now, at this very moment, it is possible that our planet's atmosphere, its weather patterns and the life that is continually developing on it, are all directly affected by ongoing bombardment from space. In that event, Earth's environment is much more complex than ever previously imagined, being determined not only from within, but at least as much from without. Perhaps all life evolved, died and was born as a result not only of volcanic, plate-tectonic, oceanic and atmospheric change, but also of destruction by the catastrophic impact with deathrocks and the spread of biological components carried on these space-born interlopers. If so, then *Homo sapiens sapiens* not only owes his existence, in part, to these extraterrestrial rogues but must also logically die as a consequence of their interaction with Earth.

One mass of rock and ice, falling through space, might destroy life where it is bountiful and seed it where it is nonexistent.

Why should this concept seem so incredible? This is a new science, a new doctrine, but it is one that must be fought for by scientists and disseminators of knowledge, as all theories of merit have been in times past. Unfortunately, this is one theory that must achieve acceptance soon, as time is against humankind.

As more discoveries of so-called near-Earth objects (NEOs) – comets, Earth-crossing asteroids and meteorites – are made, so the risk of impact from a large, hitherto undetected, object increases. *Impact Earth* investigates a trail of evidence to expose an international conspiracy of silence concealing the danger of impacts and reveals the true importance of 'deathrock' collisions in human history.

This book is a warning. As such it attempts to depict the possible nature of our world's would-be destruction in what is unhappily termed a 'factional' scenario: that is one based on scientific fact, but interpreted in a fictional way. This new form of dissemination, employed in Part II of *Impact Earth*, is growing in popularity and is one now frequently used in filmed documentary in an effort to make the intangible relevant and personal.

Each of us views our world as an unborn child might view its world: the womb is everything; the child can detect no colours, odours or sounds save garbled murmurings from outside. Baby's world seems safe and offers everything needed to survive and thrive. Yet baby is ignorant of the true world beyond its own. If its mother were to be hit by a moving vehicle, perhaps rupturing her womb, then baby's world would be shattered, its life endangered if not destroyed along with the life that begat it.

We are that child; our world the womb. Surely we should set our collective scientific and political systems to the task of genuinely determining the risk, reassessing our position in the solar system – which is more like a battlefield than the serene ether suggested by uniformitarian doctrine – and take steps to protect our world.

PART I

ONE

THE NATURE OF THE THREAT

Every year, forty tons of space rock impacts with Earth. That is an empirical fact, proven by collected samples. The problem with samples, however, is that they are prone to error. Some meteorites and asteroids (stony) are composed of material that makes them almost, visually, indistinguishable from terrestrial rocks. It seems likely that millions of tons of debris may be pummelling Earth every year – we just have to know where to look for it.

It is estimated that more than 2,000[1] Earth-crossing asteroids (ECA) exist in space, larger than a kilometre in size, and it is believed that thousands more await discovery in our solar system. Thousands more comets and meteoroids may add to this threat. A single impact by a rock the size of the Millennium Dome could devastate the surface of the globe with an explosive release of energy five times more powerful than the entire world's nuclear arsenal. On 19 May 1996, just such an object came within 280,000 miles of Earth: six hours from collision.

Humankind could have been eradicated.

This asteroid (named JA1) sailed into our system – the largest object to approach Earth, other than the moon, since records began in 1833 – and was only four days away before two astronomers (Tim Spahr and Carl Hergenrother) in Tucson, Arizona, detected it and alerted the US National Aeronautics and Space Administration (NASA). No one was prepared. Nothing could be done to prevent its approach. Yet no one was told: no public warning was given. The world's powers watched the asteroid's approach, impotent and unable to prevent the end of human civilisation. At the last moment, when it was only 400,000 miles, or seven hours, away from impact, its trajectory carried it away from our world. If it had not, then human history could have come to a violent end.

[1] Rabinowitz, D, et al, *Hazards Due to Comets and Asteroids: The Population of Earth Crossing Asteroids*, University of Arizona Press, 1994, p. 285.

On 4 November 1996, Edward Teller (b. 1908), a Hungarian-born American physicist, wrote to the then British Prime Minister, John Major. Teller distinguished himself in the 1940s and 1950s, not only contributing to the Manhattan Project (which made the first atom bomb), but also devising many of the equations necessary to produce the devastating hydrogen bomb. Teller made many enemies within the scientific community during the 40s and 50s by testifying against Robert Oppenheimer (1904–67) – the leader of the Manhattan Project. Realising that he and his team had released a dark and terrible force upon the Earth, Oppenheimer became a staunch anti-nuclear campaigner. Although Teller did offer support for Oppenheimer's loyalty to the US flag, he also questioned his character and resolve. Despite this, Teller is unquestionably a genius, one long committed to defence initiatives and a staunch advocate of the Strategic Defense Initiative (Star Wars), a programme – first introduced by the then US President Ronald Reagan in 1983 – which proposed a space-based network of antinuclear missile technologies to protect the USA.

Teller's theories led to the development in 1952 of the hydrogen bomb – a device capable of generating far more explosive force than the original 12,000 tons of TNT equivalent that the Manhattan Project's original uranium 235 and plutonium[2] (radioactive fuels derived from ores) devices were capable of. The largest hydrogen-bomb detonations can release an energy equivalent of 60 million tons of TNT; the USSR test detonated such a device in 1961. Teller's H-bomb equations are considered so important that they are still classified top secret by the US government. This is a mathematician and physicist of the highest calibre, a man whose mind has literally influenced the way late-twentieth-century history unfolded.

Despite the awesome destructive power of the devices he begat, Teller believes the greatest threat to humankind is not nuclear war, but asteroid or comet impact.

His letter to John Major PM warned:

Every few human lifetimes, there is a bombardment event like that which occurred in Siberia in 1908 (Tunguska), wiping out most life over an area of about 10,000 square miles . . . Quantitatively, the time-

[2] Atkinson, A, *The World in Our Hands*, Reader's Digest/Toucan Books, 1999.

averaged loss-of-life is comparable to that due to large floods, earthquakes and aeroplane crashes ... The advent during the last half-
century of reasonably large-scale rocket propulsion has given us the
technological means necessary to avert such impacts.

A singular warning from a singular mind. Unfortunately, this letter
and others like it to political leaders around the world have so
far gone unheeded. The future of Earth still hangs in the balance,
defenceless against dark objects hurtling through space at 70 kilometres per second (km/s)[3]: 119 times faster than Concorde's Mach
2.[4]

It is estimated, thanks to fossil remains, that two-thirds of all
animal life became extinct because of impacts from nonterrestrial
objects. Today, more than 100 ancient craters have been located on
the surface of Earth, with diameters in excess of 50 miles. They
represent only a small percentage of the large-impact craters waiting
to be discovered. The majority will be found beneath our ocean
floors – 72 per cent of the Earth's surface. These ancient scars are
slowly being brought to light thanks to new satellite technology and
more traditional geological surveys. Using computer models of the
rock beneath the sea bed, geologists have been able to produce
'gravity maps' of these incredible impact sites.

It is obvious, in the vernacular of the US military, that there is a
'clear and present danger' to the world from impact with nonterrestrial objects.

A document on the subject of impact threat – now cleared for
public release by US security and policy review authorities – was
prepared by the academic-research section of the US Department of
Defense and was presented by Colonel John M Urias and colleagues
to the US Air Force in 1996.[5] A directive issued by the chief of
staff of the US Air Force had called for an assessment of concepts,
capabilities and technologies that the USA will need to maintain an
unimpeachable airspace and to ensure that the US air forces maintain
their domination of the skies. It is clear from the tone of the report

---

[3] Morrison, D, Chapman, CR, and Slovic, S, *Hazards Due to Comets and
Asteroids: The Impact Hazard*, University of Arizona Press, 1994, p. 61.
[4] Concorde's Mach 2 is approximately 2,125 km/hour.
[5] Urias, Col. JM, et al, *Planetary Defense: Catastrophic Health Insurance
for Planet Earth* – a research paper, 1996.

that some factions of the US military believe the problem to be very serious indeed:

> Due to a lack of awareness and emphasis, the world is not socially, economically, or politically prepared to deal with the vulnerability of the EMS (Earth Moon System)-to-ECO (Earth Crossing Object) impacts and their potential consequences. Further, in terms of existing capabilities, there is currently a lack of adequate means of detection, command, control, communications, computers, and intelligence (C4I), and mitigation . . . In terms of courses of action in the event of a likely impact of an ECO, other than a nuclear option, no defensive capability exists today. However, new technologies may yield safer and more cost-effective solutions by 2025. These authors contend that the stakes are simply too high not to pursue direct and viable solutions to the ECO problem. Indeed, the survival of humanity is at stake.

The incident in May 1996, and many others like it, reflects the seriousness of the problem facing Earth. We have survived this long simply by good fortune alone. Dr Monica M Grady, head of the Petrology and Meteorics Division of Britain's Natural History Museum, shares the concerns expounded by Colonel Urias and Edward Teller.

Like many scientists, Grady accepts the definition of global catastrophe as being one that sees the destruction of one-quarter of the world's human population[6] – which means the deaths of 1,500 million people (current estimates place human population at 6 billion). This level of destruction, caused by an impactor such as a comet or asteroid, would occur not simply because of the impact itself, but the resultant disturbance to the world's atmosphere, climate and surface. An 'impact winter' might occur in which vast quantities of dust would be hurled into the atmosphere, where it might remain at high altitudes for many months. The dust may contain high levels of sulphur dioxide and nitrogen oxides, for example, if the rock it impacted with contained high levels of sulphates. The dust may therefore produce highly poisonous gases or fall back to Earth as highly corrosive acid rain. Nitric oxides are a well-known ozone-depletion mechanism – vast quantities of this material would therefore severely affect Earth's protection from high levels of harmful

[6] Morrison, D, Chapman, CR, and Slovic, S, *Hazards Due to Comets and Asteroids: The Impact Hazard*, University of Arizona Press, 1994, p. 59.

ultraviolet (UV) rays from the sun, perhaps resulting in burns, cancers and crop failure. Other elements with deleterious effects might be thrown into the atmosphere, to become aerosol, including heavy metals such as cadmium, lead and mercury.

While the dust is suspended in the atmosphere, sunlight would be reflected away from the Earth's surface back into space. The Earth's temperature could fall radically, resulting in a prolonged winter. Plants would be starved of sunlight, resulting in a breakdown in photosynthesis. They would soon begin to die and would therefore no longer be able to fulfil their dual role of cleansing the carbon dioxide from the atmosphere – producing oxygen in the process – and providing food for herbivorous animal life. Meat-eating animals depend upon herbivores for their food and so they too would suffer terrible losses. Not only would their prey begin to diminish in number, but their visual acuity, vital to all great hunting carnivores, would be undermined by dust motes and lack of light.

Impact fires would be triggered by material from the body, super-heated during its ejection and by air friction during its consequential descent back to Earth. These would spread rapidly (ejecta might be flung thousands of miles from the point of impact), throwing even more dust into the atmosphere. Although we instinctively associate fires with heat, the soot would help 'load' the atmosphere with particles which would further reduce the penetration of solar radiation to the Earth's surface: soot cooling. Omnivorous humankind, dependent as it is, to varying degrees, on both forms of food, would also suffer terrible losses during the impact winter.

The immediate effects of such an impact would be worsened by other mechanisms of destruction: vast tsunamis (oceanic waves) and shockwaves that would rip through the Earth's crust, making it move like water, followed by prolonged aftertremors and earthquakes.

Grady estimates that an energy release of 100,000 (10^5)–1,000,000 (10^6) million tonnes would cause such a catastrophe. She believes that an object capable of generating this amount of energy impacts with Earth every 0.5×10^6 (500,000) years. Her estimates are based on current counts of large impact craters on Earth (such as Chicxulub in the Yucatan) and records of environmental disruption, demonstrated by certain fossil remains and other information such as atmospheric dust found in ancient continental ice sheets.

So: 1.5×10^9 dead in 0.5×10^6 years

In other words, if this form of impact and death toll were averaged over the whole period – 500,000 years – it would be equivalent to 3,000 people dying every year (averaged over Earth's surface) for half a million years.

It is possible to calculate the risk to every human, using this approach. There is a 1-in-24,000 chance that you will be killed thanks to such an impact, at some point during an expected 70-year life span.

This level of risk is extremely high when compared with the chances of your developing new-variant Creutzfeldt-Jakob disease (CJD), better known as the human equivalent of mad-cow disease (or BSE), from infected British beef: 1-in-15-million chance,[7] during such a life span. Although the report claims that this risk assessment has 'a wide range of uncertainty', it is unlikely to amount to a factor of 625, which is the multiple needed to bring Grady's risk assessment of death from an impact down to the level of risk of infection from BSE-riddled meat.

Despite the fact that you are 625 times more likely to die from an impact, the British government, Europe and most of the world chose to ban British beef from the tables, while totally ignoring the much more pressing problem of imminent demise via a 'deathrock'.

Britain's Ministry of Defence Accident Assessment Unit, worried by the threat of impact, turned to Grady, one of Britain's most respected voices on all things meteoritic, to answer a question about the integrity of British national safety. They asked, 'What is the chance of a meteorite falling on a train carrying nuclear waste, or on a nuclear power station?' Their motives are self-evident: any such impact, however small the impactor may be, could be regionally catastrophic.

Grady based her calculations on currently available empirical evidence for the flux, or regularity of impact, of meteorites. She assumed an average of a hundred 1-tonne meteorites falling over an area of 500 million square kilometres, per year.

[7] Based on estimates published in *Assessment of Risk from Possible BSE Infectivity in Dorsal Root Ganglia*, a report produced for the Ministry of Agriculture, Fisheries and Food and the Spongiform Encephalopathy Advisory Committee. Produced by Det Norske Veritas, 1997.

so:	100 per 500×10^6 km^2 per year
i.e.	0.2×10^6 km^2 per year
therefore	there is a 1-in-5-million chance of a 1-tonne meteorite falling on any one patch of land 1 km^2 in area.

This calculation still does not relate favourably to the risks of CJD infection, yet the MOD has not taken action, whereas the Ministry of Agriculture certainly did, even to the point of banning certain cuts of meat (meat on the bone). Why such a strange disequilibrium in response to risk? Perception. The UK government believes that voters are intelligent enough to question their elected leaders' response to a perceived problem, however insignificant it may be, with the nation's food supply. But perhaps the government feels sceptical that the voter has enough information, or intelligence, to understand the potentially much greater risk of death from space. Indeed, nation-state governments seem unlikely to respond to this risk until perceptions are changed and the voter demands action, much as occurred with the threat of global warming in the 1980s and 90s. We will return to this topic later.

Understanding the nature of the threat is vital if we as a society are to develop very accurate risk assessments and possible plans of action. The science is not yet a century old: Daniel Barringer first theorised in 1905 that a crater (now known as Barringer crater) in Arizona, USA, was created by an asteroidal impact. Although this impactor was small – approximately 30 metres, weighing 60 million tonnes – it generated an explosion equivalent to 3.5 million tonnes of TNT and excavated a crater 1.2 kilometres in diameter and 50 metres deep. This first theoretical breakthrough fostered a whole new science, barely embraced by geology, physics and archaeology, even until the 1990s.

Dr Michael Baillie, a palaeoecologist from the School of Geosciences, Queen's University, Belfast, believes that there is still a great ignorance of the dangers posed by impactors, even within the archaeological community. He believes that there is still no archaeological paradigm to deal with the historical presence and influence of impacts. He recently asked for a show of hands at an archaeological conference, when he asked the audience if they knew of the impact phenomenon and its probable role in killing the dinosaurs and its relevance to human history. Only 10 per cent responded in the affirmative, indicating that there is still little cross-discipline understanding of the complexities of this issue, even within relevant scientific communities.

Perhaps this is another indication of ignorance generating bliss; but chasing false hopes that all will be well in the cosmos if we close our minds to a problem is surely tantamount to chasing a will-o'-the-wisp. Nature is seldom kind to the slothful or the apathetic.

This astonishing ignorance prevails despite the fact that one of the leading theories, continually discussed in scientific journals, to have emerged in the late-twentieth century depends upon an Earth-shattering impact by a 10–20-kilometre comet or asteroid.

This impact theory – relating to one of the most important mass extinctions ever to have occurred on Earth, where a total of 75 per cent of all life, animal and vegetable, became extinct at the boundary between the Cretaceous and Tertiary periods of geological history, 65 million years ago – has set alight the minds of film-makers and theoreticians alike. The concept has caught the public imagination and fuelled endless speculative fiction, and yet as a whole the scientific community has been uniformly slow in understanding its wider significance across all disciplines of physical sciences, from astronomy to geology, and of the social sciences of economics and history. It is my belief that the process is only just beginning. The science is embryonic, yet evidence gathered from a multitude of scientific disciplines is drawing a very grim picture of all our futures.

This discussion is all very well, but it still seems a remote threat – somehow removed from our Earthly troubles: political in-fighting, threat of war, United Nations peacekeeping missions and our own domestic wrangles. Yet, if we could stop the magic life-roundabout and get off for one moment, step back from our work-dependent rush-hour-scarred lives, we might look at ourselves as a stranger from another culture could. If we were lucky we might see homes filled with love, and of course joy and sadness. This is something special, and found in different ways throughout the animal kingdom: there is always a 'nest' and often loved ones within. If we could look at all this – our homes, partners, children, all of our achievements – with the knowledge that in a few moments it will all end, then perhaps we could begin to understand what such an impact might mean.

Society is important and learning, art and culture are all defining characteristics of *Homo sapiens sapiens*. Yet these facts fade like snuffed candles when viewed next to the brilliant white light of our

Geological Timeline: 550 Million Years BC to Present Day

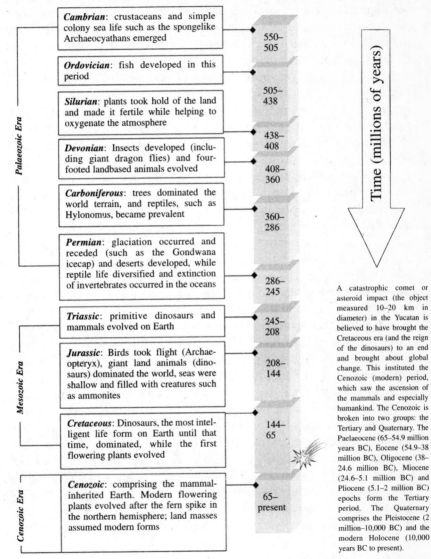

Cambrian: crustaceans and simple colony sea life such as the spongelike Archaeocyathans emerged

550–505

Ordovician: fish developed in this period

505–438

Silurian: plants took hold of the land and made it fertile while helping to oxygenate the atmosphere

438–408

Devonian: Insects developed (including giant dragon flies) and four-footed landbased animals evolved

408–360

Carboniferous: trees dominated the world terrain, and reptiles, such as Hylonomus, became prevalent

360–286

Permian: glaciation occurred and receded (such as the Gondwana icecap) and deserts developed, while reptile life diversified and extinction of invertebrates occurred in the oceans

286–245

Triassic: primitive dinosaurs and mammals evolved on Earth

245–208

Jurassic: Birds took flight (Archaeopteryx), giant land animals (dinosaurs) dominated the world, seas were shallow and filled with creatures such as ammonites

208–144

Cretaceous: Dinosaurs, the most intelligent life form on Earth until that time, dominated, while the first flowering plants evolved

144–65

Cenozoic: comprising the mammal-inherited Earth. Modern flowering plants evolved after the fern spike in the northern hemisphere; land masses assumed modern forms

65–present

Palaeozoic Era

Mesozoic Era

Cenozoic Era

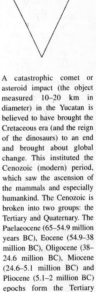

Time (millions of years)

A catastrophic comet or asteroid impact (the object measured 10–20 km in diameter) in the Yucatan is believed to have brought the Cretaceous era (and the reign of the dinosaurs) to an end and brought about global change. This instituted the Cenozoic (modern) period, which saw the ascension of the mammals and especially humankind. The Cenozoic is broken into two groups: the Tertiary and Quaternary. The Paelaeocene (65–54.9 million years BC), Eocene (54.9–38 million BC), Oligocene (38–24.6 million BC), Miocene (24.6–5.1 million BC) and Pliocene (5.1–2 million BC) epochs form the Tertiary period. The Quaternary comprises the Pleistocene (2 million–10,000 BC) and the modern Holocene (10,000 years BC to present).

11

families. Most of us lose loved ones at some point in our lives, be it a parent or grandparent, partner or friend. This terrible feeling of loss is common in the wider animal kingdom too: nonhuman creatures can grieve for a lost mate. Bearing these naked feelings in mind we might be able to understand the terrible dread, unbridled fear and grief that might have gripped the Earth when such an impact did unleash a catastrophic maelstrom 65 million years ago.

Yes, the creatures then inhabiting Earth were almost unimaginably different from us. Yet those differences may not be so great as we might at first expect. We know dinosaurs nestled their young, invariably laid eggs in nests and hunted for food much as a bird might today. Although a popular theory by Professor John Ostrom[8] (Yale University), suggesting that birds are descendants of dinosaurs, has been largely dismissed by the embryological work of Alan Feduccia[9] – which proved that bird wings are derived from the second, third and fourth digits of their vestigial 'hand', whereas the 'hands' of theropod dinosaurs were derived from the first, second and third digits – we do know that many dinosaurs nested and behaved as birds might. The herbivorous duck-billed *Maiasaura* genus, is a dinosaur that is well known for its parenting skills: nests filled with young and a dozen or more eggs have been found in Montana. The finds indicate that the young were fed by their parents, in this nursery environment, and nurtured until they could begin to forage, when they would in turn bring food back to the clutch to feed another generation of brothers and sisters.

It is surely inconceivable that creatures that evolved from a line that had inhabited Earth for 150 million years were without feeling, whether it be parental instinct or some deeper emotion. They must have developed some community, as all animals on modern Earth have, in order to survive. Indeed fossil finds have indicated that many of these creatures moved in herds and probably lived within organisational structures approximate to those seen within life on Earth today. The *Maiasaura* dinosaurs' parenting could indicate that dinosaurs had a true flocking 'group' mentality. We are all herd creatures, even humans – why else would we seek companionship, discuss common problems and look to banks of collectively derived

[8] Ostrom, JH, *Dinosaurs*, Carolina Biological, 2nd edition, 1984.
[9] Feduccia, A, and Burke, AC, 'Developmental Patterns and the Identification of Homologies of the Avian Hand', *Science* 278:666, 1997.

knowledge for solutions to our problems? What is this sort of behaviour but a manifestation of the human flock?

Perhaps we can sympathise, then, with this most ancient creature, as it looked upon its children for the last time, possibly instinctively comprehending that they were about to perish. Death would have come to them as it comes to all of us: unwelcome, terrible and without mercy.

Although the killing mechanism for the extinction of 75 per cent of all life on Earth, as occurred at the junction between the so-called Cretaceous and Tertiary periods of geological history (referred to as the K/T boundary), is still hotly debated, evidence weighs heavily in favour of an impact with a 15-kilometre deathrock. If so, then the impact literally brought the worst concepts of biblical Armageddon to reality.

What happened?

Without warning, a vast impact: an energy burst equivalent to 100 million million tonnes of TNT. To put this impact in perspective, if humankind wanted to simulate the event as an experiment, perhaps on the surface of Mars, using a cache of post-World War Two, Hiroshima-like, uranium-fuelled atomic bombs, each with an equivalent yield of 12,000 tonnes of TNT, we would need to detonate 8,333 million bombs at the same instant to release as much energy as the K/T deathrock.

A dinosaur standing in the centre of North America would have seen a brilliant flash, brighter than the sun, overhead. Yet no sound would have heralded its coming – nor would sound warn of a tremendous shockwave and shower of high-temperature debris travelling at several times the speed of sound. The incredible mortal boom would eventually arrive, being felt as well as heard, as the Earth rang like a bell in response to the 10–15-kilometre deathrock's power. Earthquakes would have radiated around the world, the seismic activity prompting volcanic eruptions.

The dinosaur and its family would probably be running, taking to their heels to escape the falling death, in vain. Fifteen minutes after impact, death finally arrived: a down-range vapour trail travelling at ballistic speeds covered North America – raining down on the hapless family, killing them with a myriad bullets of rock.

The impact, 2,000 kilometres away, in an an area that would come to be known as the Yucatan Peninsula, excavated a 100-

kilometre-deep hole in the bedrock, generating a 180-kilometre-wide cavity. A fireball, a living wall of flame, consumed North America and crept around the northern hemisphere, shrouding the planet in a ball of fire and worsening the great airborne dust cloud. The bodies of the family would have been consumed by the raging heat, perhaps as hot as 2,000 degrees Celsius – the very air afire. Anything and everything flammable was laid to waste by flame.

The energy of the impact was intense, yet much of it would have been directed into the atmosphere, owing to the angle of impact, thus worsening the disaster because the Yucatan Peninsular bedrock had a rich calcium sulphate (anhydrite) content, creating sulphur dioxide, a deadly gas. Worse still, the gas combined with water in the stratosphere and formed a yellow haze, and this, combined with the tremendous dust thrown into the atmosphere from ground zero, resulted in an effective shield from the sun. The skies fell dark. Over the period of a year, global temperatures fell by 2 to 10 degrees Celsius – a catastrophic figure. The result: a frozen world.

Vegetation that had not died during the fireball died slowly, thanks to nitric-acid rain and lack of sunlight: all photosynthesis ended. All large animals, over approximately 25 kilograms in weight, died. Only the foraging animals, such as small mammals that lived off insects and creatures that in turn lived on dead tissue and decomposing vegetation, survived. Earth had become a home without a family, a great cache of decomposing corpses.

The light had gone. The oceans' phytoplankton, dependent on the sunlight for its food, died. Sea life that depended on the phytoplankton for food, starved. The food chain in the seas collapsed, killing much of the population, such as the ammonite, leaving only those creatures that foraged in the sea bed, such as the nautilus, to survive. The oceans were decimated within six months.

Just as the world had frozen thanks to sulphur cooling, now it entered a radical period of global warming thanks to the aerosol CO_2. For the next century, the impact's CO_2 legacy increased global temperatures by 10 degrees.

Seventy-five per cent of all species passed from Earth: extinct.

Is this description accurate? It depends upon which impact theory one believes. If the strike was a direct hit at right angles to Earth then no, the above description will not work, but, if it was a low-angled oblique strike, approximately 30 degrees,[10] then unquestionably the world was shaken as described. The evidence now weighs in favour of an oblique strike. But finding that evidence required a cache of tools from the sciences of geology, physics and palaeobiology as well as a good deal of rational thinking and detective work.

In June 1979, a seminal scientific paper[11] was accepted, and published five months later, without herald or acclaim, by the respected British journal *Nature*. Titled *A Theory of Terrestrial Catastrophism*, the paper, by the British astronomers William Napier and Victor Clube, presented for the first time a rationale for a possible impact-extinction mechanism. Within seven months, a rival group, originating in the USA, headed by the Nobel Prize winner Luis Alvarez (1911–88)[12], and his son Walter (b. 1940), linked this form of extinction with the greatest mystery of geology: the extinction of the dinosaurs.[13]

The Alvarez team, composed of Rich Miller (a physicist), Frank Aasaro (nuclear chemist) and David Cudaback (astronomer), was drawn to solve a mystery unearthed by Walter Alvarez during research in Gubbio, Italy. The Bottacione Gorge near Gubbio, created by sedimentary deposits, is a rich source of information about the Cretaceous period in Earth's history. It was here that Walter Alvarez realised that there was an extraordinary layer of clay that was apparently deposited conterminous with probably the most dramatic extinction in prehistory – 65 million years ago. The clay layer actually delineated the boundary between the Cretaceous and Tertiary periods. Researching the layer's origins, they were amazed to discover that the deposits were common to other areas – in a drilled core from beneath the Pacific sea bed, in the Raton Basin on the border of Colorado and northern New Mexico – and by the mid-1980s it was apparent that the layer was to be found across the globe.

[10] Schultz, P, 'The Deadly Oblique Angle: North America Hit Hard by an Asteroid Strike in the Yucutan 65 Million Years Ago', Brown University, RI, PR: 96–041, 1996.
[11] Napier, WM, and Clube, SVM, 'A Theory of Terrestrial Catastrophism', *Nature*, Vol. 282, Nov. 1979.
[12] Luis Alvarez was awarded the Nobel Peace Prize for his high-energy-physics research in 1968.
[13] Alvarez, W, *T. Rex and the Crater of Doom*, Penguin, 1998, p. 80.

Searching for a causal nexus to this layer and an apparent mass extinction revealed by fossil records – especially the abundance of forams (small marine life) before the K/T boundary and their rapid disappearance thereafter – Walter turned to his father Luis. Probably influenced by Napier and Clube's paper on terrestrial catastrophism, published in 1979, Luis suggested a catastrophic impact with a vast asteroid or comet as the origin of the clay layer. After several false starts, they found a method of determining the origin of the clay layer. If the mass extinction had been caused by a catastrophic impact there would be a signature or fingerprint.

Iridium is a very rare element. Geological theory speculates that iridium, palladium, ruthenium and other rare minerals were subducted, carried, to the Earth's core, when the Earth was young and its surface liquid and molten. Despite this rarity on Earth's surface, it is particularly common in stony asteroids and meteorites and the clay layer revealed a 9-parts-per-billion[14] concentration of iridium – normal earth levels would be in the order of 0.1 parts per billion. They had already abandoned thoughts of a supernova[15] cause for the mass extinction on Earth, as the clay layer would have been rich in the radioactive plutonium 244 isotope and yet it was noticeably absent, despite the fact that it has a half-life of approximately 85 million years. With the discovery of the iridium layer, the Alvarez team had their deathrock signature.

They published their work in the June 1980 edition of America's foremost scientific journal, *Science*. There was uproar and an immediate backlash from the uniformitarian school, of which, as a geologist, Walter Alvarez had formerly been a part. As in all things, humankind dislikes change. History shows that, whenever a new concept is presented to a community of like-minded individuals, derision and scorn will almost inevitably be poured upon the originators: Galileo Galilei could, were he alive in the 1980s, have sympathised with the Alvarez group.

Nevertheless, the theory gathered converts and these disciples of the new doctrine began to hunt for an answer to the uniformitarians'

---

[14] Alvarez, W, *T. Rex and the Crater of Doom*, Penguin, 1998, p. 69.
[15] The explosive death of a star, a moment in which it generates more visible light than thousands of millions of stars, can result in the bombardment of other solar systems and planets with rare elements and also releases matter vital to the formation of other worlds.

demands for more proof. They wanted to see the smoking gun: a vast impact crater.

Finally, in 1990, ten years after the initial publication, one man pieced together a trail of evidence, itself ten years old, to conclude that the crater of that supreme smoking gun was buried beneath what has become the Yucatan Peninsula, at a place known as Chicxulub.

Alan Hildebrand of the Canadian Geological Survey employed a mathematical model of the K/T impact site, built from data gathered by oil-hunting geophysicists working for Mexico's national oil company Petrolenus Mexicanus in the early 1980s. The geophysicists employed a number of techniques, particularly seismic surveys and core samples. The seismic readings were made by detonating explosive charges near the sea bed and measuring the time the explosive reverberation took to return to the point of origin: analogous to radar. The core samples were taken during exploratory drilling in the sea bed. Hildebrand was the first to use the geophysics data produced in this way to hunt for, and find, the Chicxulub crater – possibly the largest of all Earth's impact scars.

If Earth is indeed threatened by another approaching deathrock – perhaps tomorrow, or next year, or even thirty years from now – Hildebrand, the Alvarez team, Napier and Clube may be seen as playing one of the most vital roles in human history: the saving of our civilisation. How? They dared to raise a bold new theory, and search out tangible evidence, to prove that a frightening killing mechanism, hitherto ignored, had existed and exists today, its threat to humanity far from extinct.

Although there is still some contention regarding this K/T impact scenario, the detractors are now in the minority and look increasingly like radicals. Monica Grady[16] states:

> There's no doubt that there is a crater, in the Gulf of Mexico at Chicxulub. It's huge. There's been a lot of debate between the people who have done the gravity work about whether the crater is 180 or 300 kilometres across. There is an enormous crater and the age of the crater is the same as the end of the Cretaceous – there's little doubt about that at all. There's very little doubt that the environmental consequences of an impact like that would be absolutely catastrophic.

[16] Grady, M, interview, 1998.

With the anhydrite and sulphides in the bedrock the impact could not have been in a worse site.

The chemicals thrown into the atmosphere had an extremely detrimental effect on the world's environment. The sulphides in particular would have produced a sulphur-cooling effect – in essence a barrier to solar radiation, throwing vital heat energy back out into space and away from the Earth's surface; the result was a global freeze-up. The tidal waves and resultant aftereffects of the impact would have been catastrophic. Vast quantities of debris were thrown into the atmosphere including large quantities of iridium. Grady continued:

> There is no doubt that the crater was created by an impacting bolide. The impact would have had horrendous consequences in terms of tidal waves. Energy of the impact would have radiated back to Earth, creating firestorms, swinging temperature changes – very hot then very cold and then warm again – changes possibly in the composition of the sea: probably a pH change. The catastrophe would have been global – the iridium layer is evident all over the world, in New Zealand, the Americas, Europe, Scandinavia, Africa. The layer has been found right across the world. There's no doubt about that. We have a cause and we can certainly see an effect. But the debate ensues about whether the environmental changes were sufficient to cause the global catastrophe: the loss of species.

There lay the crux of their argument: where to find probably the biggest hole on Earth? Presumably those that doubt the global effects of an impact large enough to generate a crater that is 180–300 kilometres across also fail to accept the data produced during computer simulations and scaled-down physical experiments. The uniformitarian view is one that has gripped science for so long that its tame old dogs find it hard to learn new tricks. This is understandable – why should a scientist abandon his or her doctrine when a new theory emerges? After about twenty years of study, however, I have to ask why the uniformitarians cling to their stance in the face of such overwhelming evidence.

In North America a great upswell in the population of the fern occurred in the million years that followed the K/T impact. In geological terms, that can be measured to around 10 centimetres of rock above the Alvarez iridium layer. More than a quarter of a million fossils extracted from over 300 sites in North America demonstrate high quantities of fern pollen, to the extent that 90 per cent of all flora appears to have been composed of ferns. The fern dominated the wrecked landscape, as the 65 discernible fossilised species found prior to the K/T impact had been reduced to only eight species post-K/T. The world was, in essence, new again, beginning the process of rebirth.

This fern upswing is a good indicator that the impact was oblique, as the fern spike occurred only in the northern hemisphere, and, although the whole world was affected by the impact, the pattern of most severe damage was in the northern hemisphere and especially in North America – following the direction of the impactor's trajectory. The impactor hammered the surface at a low angle, approximately 30 degrees, throwing vast quantities of energy into the atmosphere, which rolled around the Earth devastating the northern lands in particular. Had the impactor hit the Earth's crust at a 90-degree (perpendicular) angle, much less of the energy released would have perturbed the atmosphere. The consequences of a perpendicular impact would be more earthquakes and seismic activity because the energy transferred at impact is directed into the earth. The horseshoe shape of the Chicxulub crater itself, as opposed to a simple circular crater, is probably therefore a result of a low-angle strike.

The discovery of shocked quartz – quartz crystal shot through with shock faults and small glass spherules (tektites) created at the moment of impact – and impact rubble mostly concentrated at Chicxulub and to a lesser extent at corresponding sites around the world, further proved that impact had occurred, and that it was located at Chicxulub. Sedimentary rock, deposited by the vast tsunami, was also found across the globe, concentrated again in the Chicxulub vicinity. These geological footprints are unique, occurring at the same time only as a result of an impact.

Geophysics and the fossil record are but two of the tools used to determine the regularity and severity of impacts. As the investigation of past impacts gathers pace, in an effort to more accurately assess the risk to our society from future impacts, tools from other disciplines are being increasingly employed: dendrochronology, ice cores, astronomy, mathematical and computer simulations of impacts, and even the study of mythology and historical documents.

Dendrochronologists set out to build a year-by-year record of average growth for a specific type of flora with a long life span:

trees. By examining the ring patterns produced by successive years of growth within the tree trunk, and cross-referencing these with other samples taken from the same location and from a number of other sites, the dendrochronologist is able to piece together a chronology dating as far back as the earliest record will afford. This done, he then attempts to use local wood found in buildings within or near the sample area and use the ring patterns in the wood to further extend the chronology back through time.

Tree-ring chronologies for Europe and America now extend as far as the thirteenth century AD. This so-called 'master-chronology' allows buildings and other structures to be dated. Recently, however, it has become a valuable tool in assessing the possibility that impacts have been far more frequent, and dangerous, than ever previously suspected. Because tree rings reveal poor years of growth as well as the good, it becomes evident when there was an environmental downturn – resulting in poor growth. If these findings are then correlated with the products of other research methods, such as ice cores and astronomical data, it is possible to determine when, in human history, specific regions of the Earth have suffered terrible climatic perturbations. As we shall see in Chapter 2, many such events appear to have occurred since 2345 BC[17] and have continued to the present day.

Ice-core records, such as those used in geophysics analysis by oil prospectors, are drilled from stable ice sheets, such as in Antarctica and Greenland, which date back over 110,000 years. These records are particularly useful, as not only do they record climatic fluctuations via changes in ice thickness and renewal of the ice sheet over time, but the ice itself stores chemicals inherent in the atmosphere at any given moment. By studying ice cores, researchers are able to measure calcium, chloride, potassium, magnesium, sodium and sulphate levels in any given period. Similarly, they can examine the ice for traces of rare materials, such as iridium. Where a volcano has an often quite specific sulphur signature in the ice, asteroids often leave iridium fingerprints. The Greenland Ice Sheet Project 2 (GISP2), begun in 1993 and completed in 1998 and funded by the National Science Foundation of the USA, has enabled researchers to build up a good picture of climate, human response to the changing arctic

[17] Bailey, MGL, *Evidence for Climatic Deterioration in the 12th and 17th Centuries BC*, Oetker-Vogues-Verlag, 1998, p. 1.

environment, and climatic interactions between the northern and southern hemispheres.

Although the ice-core data is still being assessed and the picture is far from complete, owing to the regional nature of the sampling, the GISP2 may help to reveal the regularity of impacts in history when studied alongside the findings of dendrochronologists, geologists and astronomers.

Historical records also provide vital clues: where there is no hard scientific reporting of phenomena in early human history, records by the literate and educated may offer insight. As we shall see, there are many fascinating accounts of what could only be interpreted as impact phenomena in history.

Although this cross-discipline form of study is rare, it is beginning to be embraced as the only way to develop coherent theories about how humanity and its environment have been and will be affected by impact events. There is so much information available in all fields that, if they are cross-referenced and correlated, a true picture can emerge – and is beginning to do so.

Unfortunately, very few scientists are actually studying the phenomena in any polymathic way and there is a danger that the study of times past may draw too much of their attention, with too little attention being focused on tomorrow. Indeed, worry about the future occurrence of dangerous objects has been left, thus far, largely to the world's astronomers.

It is clear, then, that we must not be lured into seeing this as an extinct threat, or be mesmerised by the threat from larger impactors. The type of asteroid or comet that caused the K/T extinction was so vast, 10–20 kilometres in diameter, that a similar occurence is unlikely to happen more than once every 100 million years. Unlikely, but not impossible.

The primary threat that must be focused upon, and that is increasingly seen to trouble the US military and space industry, is that of the deathrock in the size range of 300 metres to 2 kilometres. The risk from these objects is extremely real and can produce catastrophic results: a 1-kilometre asteroid travelling at an average speed of 50 km/s will impact with Earth with an explosive force equivalent to 30,000 megatonnes of TNT. That is the equivalent of 2.5 million Hiroshima-sized bombs.

Perhaps even more disturbingly, smaller impactors of the 50-to-100-metre range can generate an H-bomb-sized explosion: the equivalent of approximately 10–20 megatonnes of TNT. On 30 June 1908 just such an object exploded over Earth. It is likely that this deathrock was a fragment from a vast comet, probably debris from the Beta-Taurid complex of asteroids created by the break-up of the short-period Comet Encke, named after its discoverer the German astron-omer Johann Franz Encke (1791–1865), which returns to Earth once every four years. Earth travels through this complex of cometary debris, which orbits at an angle of five degrees to the ecliptic, in the months of June and November of every year. The Tunguska blast occurred during the June crossing in 1908.

The blast destroyed a vast swathe of Siberia. This spectacular blast – the largest known event in the twentieth century – occurred before the cometary fragment reached the ground, at a height of approxi-mately 8 kilometres. Because its angle of incursion was rather oblique, only 15 degrees, the resultant shockwave, shaped rather like two butterfly wings, burnt, defoliated and felled trees, and killed thousands of reindeer, across an area of approximately 2,000 square kilometres. The Tunguska forest was set aflame even to a distance of 15 kilometres.

The local Inuit tribesmen and -women, simple people with folklore traditions barely touched by the twentieth century even today, felt sure that the world had ended. The blast's shockwave travelled around the globe and registered on seismographs, ordinarily used to detect earthquakes, as far away as Belgium. Games of cricket were played in London under an incandescent night sky that was said to be so bright as to allow a newspaper to be read at midnight.

Interestingly, the Tunguska blast did produce a global scattering of iridium, which was discovered in a Greenland ice core during the GISP2 project. Though at much lower levels than that deposited by the K/T impact, this iridium anomaly still represented an eighteenfold increase on naturally occurring iridium levels.[18]

Clearly the Tunguska event could have resulted in a major catas-trophe. As Dr Jasper Wall, director of Britain's Royal Greenwich Observatory, Cambridge, pointed out to me, the Tunguska blast could have had far more serious consequences, indeed would have changed history, had it occurred at a slightly different time.

'Had Earth moved for another three hours or so before the impact

---

[18] 'Reading the Fine Print of Climate History', National Science Foundation, PR: 1994–56.

occurred', he says, 'the target would not have been a barely populated corner of Siberia, but Moscow City itself. Ten million people would have died.'[19]

Clearly a Tunguska-like blast occurring over any modern city would result in a catastrophe of almost unimaginable proportions. These events appear to occur once every 30–100 years, though still smaller incidents can occur once every decade. As we will see in Chapter 4, just such an impact caused widespread devastation in Brazil in 1931.

One sobering thought is that there are potentially millions of objects the size of a cricket ball that could penetrate Earth's atmosphere and strike our planet with an energy yield equivalent to a Hiroshima-sized atomic bomb, if one were to impact at high speed. The technology needed to detect and deflect such a small high-speed object simply does not exist. Although I believe we should focus our attentions on objects in the range of 30 metres to 2 kilometres, it is not simply because they pose the greatest threat, but because we are utterly technologically impotent to meet the threat posed by objects smaller than 30 metres. This is quite a threat: Hiroshima's 12-kilotonne explosion killed and seriously injured 50,000 people. Nevertheless, although such a death toll would be very high, we must regard such an eventuality as benign, in global terms. Indeed some 250 atomic-bomb-sized detonations due to comets and asteroids have been registered by the USA's nuclear-detection system and spy satellites over the course of one decade. Thankfully these explosions have all been at relatively high altitudes, but, with an average frequency of two a month, it seems more than likely that a cricket-ball-sized object may one day pose a threat.

Dr Jasper Wall puts this risk in perspective: 'One in twenty-four thousand is just about the same level of risk that you face when you travel by air,' he told me. 'Don't tell me that we have not all experienced worries about dying in an aircraft accident every time we get on an aeroplane. Consider the amount of money that is spent every year on aircraft safety and regulations. Whereas there is nothing spent on the impact threat, yet the time-averaged risk is virtually the same.'

Many military contractors and US government departments, including Lockheed Martin and NASA, have become extremely con-

[19] Wall, J, Royal Greenwich Observatory, Cambridge, interview, 1998.

cerned by the potential impact threat. Lockheed Martin, one of the USA's leading corporations, specialising in telecommunications, space-launch vehicles, electronics and weapons research, used its energy laboratory, known as Sandia, to research the impact threat. Based in Albuquerque and Livermore, California, Sandia is an organisation fostered by the US Department of Energy and operated by Lockheed Martin, which pursues research-and-development programmes relating to weapons, energy and environmental technologies. Using energy equations, similar to those developed by Edward Teller while he was creating the hydrogen bomb, which take into account environmental and atmospheric factors, Sandia's team of scientists were able to create virtual-reality simulations of a catastrophic impact.

With decades of experience in shock physics, and in the use of complex computer software, the computational physics and mechanics department, led by David Crawford and Arthurine Breckenridge, were able to effectively model the impact of a 1.4-kilometre-diameter comet, weighing approximately 1,000 million tonnes and travelling at 60 kilometres per second, as it strikes the Atlantic Ocean 25 miles south of New York.

The key components in the success of these simulations, apart from the human factor, are the world's most powerful supercomputers: the teraflops. They performed 54-million-cell calculations using parallel processing: thousands of computing tasks were assigned to 1,500 separate processors operating on up to one trillion mathematical operations per second for 48 hours. When these tasks were accomplished, they were reassembled within the supercomputer. Their conclusions are quite simply frightening.

The comet impacts with the ocean and instantly vaporises 300–500 cubic kilometres of water, forming a transient cavity in the ocean. An equivalent to 300 gigatons of TNT energy is released at that moment – ten times the explosive power of the entire world's stock of nuclear weapons.

Five seconds after the comet hits the ocean, a tremendous impact plume – many times the size of New York – composed of superheated debris, earth and water smothers much of Long Island.

Eleven seconds after impact the New York shoreline is swamped by superheated steam and ejecta-debris. A good amount of the ejected-debris cloud has penetrated Earth's atmosphere on suborbital trajectories. The heat generated at the impact point is in excess of

5,000 degrees Celsius. New York soon succumbs to the falling molten ejecta and the heat generated by the impact incinerates cities and forests instantaneously. The global debris cloud lowers temperatures worldwide. Then there are snowstorms and major temperature drops that last several weeks.

The Sandia simulation borders on what is described as a global catastrophe: one that is expected to cost the Earth one-quarter of all its human inhabitants. Make no mistake: this is no computer game. Sandia's team are far from new to this sort of calculation, earning great respect for their simulations of a real impact (albeit on Jupiter) in 1994, completed before the real event, and mimicking the actual effects with a fine degree of accuracy. David Crawford and his colleague Mark Boslough were the team responsible for the 1994 simulations, which modelled the awesome destructive force of comet Shoemaker-Levy 9 as it plunged into Jupiter's surface.

'A lot of major breakthroughs in science are going to come through these kinds of calculations,' Crawford says. 'It's almost like doing an experiment – one you could never do. One you would never want to do.'[20]

With the accuracy of Sandia's computations proven by this spectacular piece of prediction it is clear that the impact of a small comet in the ocean would be akin to the worst concept of nuclear war. A concept that obsessed whole nations during the Cold War, yet one that is dwarfed by the magnitude of this threat.

We as a society now face a difficult question: is Earth and its human population worth saving? Society is composed of groups of people, invariably beholden to a large corporation or employer, each in turn dependent upon the world stock markets. Perhaps, therefore, an economic view of Earth will enable us to assess its worth.

The land mass and oceans of the Earth have been assigned a value of $86.75 trillion[21] (£54.22 trillion); Earth's human life has a value based on its ability to pay and is assumed to amount to $1,900 trillion (£1,187.5 trillion).[22] The total value of Earth then is estimated to be $1,986.75 trillion (£1,241.72 trillion).

Make no mistake: were even a small impact of the Tunguska type

[20] Sandia National Laboratory, PR: 5 May 1998.
[21] A trillion is assumed to be 10^{18}.
[22] Salt, JE, *Catastrophic Risk: An Insurance Perspective*, Loss Prevention Council, 1998.

to devastate a major city, the world's economy would be thrown into turmoil, possibly even a crash. As the disaster suffered by the Asian tiger-economies, and that of Russia's crumpling economic infrastructure, ably demonstrated in 1998, no one country is immune to shocks in another. The world markets are now so inextricably linked, international trade so important, that they cannot help but import inflation and suffer from falling demand and consumer confidence: the system might collapse like a house of cards, as it did following the Wall Street Crash of 1929. The Great Depression, precipitated by the crash of 1929, saw industrial stocks fall by as much as 80 per cent, while prices for agricultural products fell by 50 per cent in the US between 1929 and 1932. Hyperinflation, where prices rise by thousands of percentage points per year, gripped Germany, resulting in terrible hardship, and fostered the rise of the Nazi party. The rest of Europe suffered, as America suffered, mass unemployment, misery and anguish. The passage of seventy years changes little except that, if anything, the world economy is even more fragile to such shocks, owing to its increasing dependence on international trade.

In these calculating terms, the impact threat is frightening enough, yet such losses pale when compared with human life, especially the lives of those we know and love. Perhaps in the new millennium we will achieve the goal of charting every object in the heavens and determining some way of stopping the deathrock that, as the laws of probability tell us, bears our name.

The report by Colonel Urias and his colleagues reminds us of the true importance of this problem, not just as a technical challenge but as perhaps the greatest life-saving venture ever considered by humankind.

> If some day in the future we discover well in advance that an asteroid that is big enough to cause a mass extinction is going to hit the Earth, and then we alter the course of that asteroid so that it does not hit us, it will be one of the most important accomplishments in all human history.[23]

The clock is ticking.

[23] Urias, Col. JM, et al, *Planetary Defense: Catastrophic Health Insurance for Planet Earth* – a research paper, Chapter 1, 1996.

TWO

ENGINE OF EVOLUTION, DESTROYER OF HISTORY

Life has existed on Earth for approximately three and a half billion years. Yet if that span were represented by a yearly calendar, with life beginning on 1 January, human beings would not appear until one minute to midnight on 31 December. The history of humankind, therefore, represents only a tiny fraction of the Earth's past, yet even now the greatest conceivable threat may terminate our shorthold tenancy on this world.

The late Eugene (Gene) Shoemaker (1928–97), one of America's foremost geologists and astronomers and a founding father of planetary science, was motivated in his mission to protect our society by the thought that much that was and is alive on Earth has been influenced by the impact phenomenon. Shoemaker, a man skilled in disseminating complex theory to the layperson, has probably done more than anyone to shift this phenomenon from the recherché domain of the scientific cloister to the main stage. He was of course helped, in part, by the vast demonstration of his concerns – the Jupiter collision – which occurred only eighteen months after he and his wife Carol, with his colleague David Levy, discovered in 1994 a column of comet fragments that became known as Shoemaker-Levy 9.

Shoemaker established and, from 1961 to 1966, acted as director of the US Geological Survey's (Flagstaff, Arizona) astrogeology department, as well as leading the geology field-investigations team for the first Apollo lunar landings (1965–70). He also chaired a National Aeronautics and Space Administration (NASA) working group on the concept of surveying near-Earth objects. So it is hardly surprising that his was a voice that commanded a great deal of attention within the scientific and political communities.

'My own hunch is that we will find that the impact of large objects, probably chiefly comets, has profoundly influenced the evolution of life,' he said in a TV documentary.[1]

This notion that life has been manipulated and steered, and that history has been altered, by the impact phenomenon has been echoed by many other leading scientists, some even going as far as to state that all life on Earth was actually seeded by cometary impact – a theory known as Panspermia. This concept is so radical that it has yet to reach the respectability of the Alvarez work and yet there is a great deal of circumstantial evidence to render the theory workable, if not immediately acceptable. Certainly a concept that will be increasingly studied in the future, Panspermia was first suggested in 1981 by Sir Fred Hoyle (b. 1915) and Chandra Wickramasinghe. Hoyle's career has been as remarkable as Shoemaker's, though a good deal more controversial. He became Professor of Mathematics at Cambridge University in 1945 and remained in that post for a decade, later appointed Plumian Professor of Astronomy and Experimental Philosophy (1958–72) and finally Professor of Astronomy at the Royal Institution of Great Britain in 1969.

Hoyle was knighted for his contribution to science in 1972 and has persisted with radical theories that have typified his career: Panspermia may well rank as the most shocking and possibly the most important. Developed in association with his former student Chandra Wickramasinghe – currently Professor of Applied Mathematics and Astronomy at the Cardiff-based University of Wales – one of the world's leading scientific authors (over 20 books and 250 papers) – the Panspermia hypothesis suggests that life did not in fact originate on Earth, but in the vast reaches of space.

Creationists – religious groups by definition – have as difficult a time digesting this theory as the traditional evolutionists. Nevertheless, the processes expressed in Newtonian science demand that this new theory must be given equitable consideration alongside those that posit a terrestrial origin of life. Any and all who would damn this theory, simply because of its newness and unabashed originality of thought, are also damning the process of good science. Good science is the process of study, hypothesis, experimentation and hopefully arrival at proof. The panspermia hypothesis has a good deal more hard physical evidence than the multifarious theories regarding the great omnipotent one's intervention on Earth – despite the fact

[1] *Horizon*: 'Crater of Doom', BBC Television, 1997.

that God is not only currently winning the argument, but also that he/she/it has supporters that number in the billions. But then he's been around a lot longer than Sir Fred. However, with a lot more analysis, we might just find ourselves considering Sir Fred Hoyle and Chandra Wickramasinghe as having the best insight into creation. Indeed one might argue that they have a better understanding of God's great DIY project than the Vatican; they would not be the first, as Galileo and Copernicus would tell you, if they could.

Their theory is elegant in its simplicity, the evidence clear and remarkable. Comets may not be merely dirty snowballs venting gas as they approach the sun but reservoirs of organic material, perhaps even living bacterial or viral organisms, suspended in a frozen or quiescent state. Formed at the beginning of time in the vast molecular clouds within our galaxy, these so-called 'biogenic comets' would have had, at one time, a liquid core – possibly amino acids and compounds of carbon, hydrogen and oxygen – and much of this material would have been processed, developing complex associations: biogenic processing. These liquid cores would later refreeze, throwing the trapped organisms into a dormant state.

These dormant organisms could then travel through the vast distances of interstellar space, over periods of time that are beyond human comprehension, perhaps billions of years. Inevitably, some of these comets came to an end, impacting with our own globe, their 'cargo' thrown to Earth's winds, 'seeding' the planet with components vital to the creation of life and, in some instances, possibly bringing living organisms to rest on Earth.

Sceptics may find it hard to voice their opinions in light of the fact that millions of 'dormant', ancient, bacterial species have been isolated in samples of Earth origin. The possibility exists, therefore, that life could be carried in comet nuclei. Biologists have come to recognise that bacterial cells, dry-frozen to temperatures approaching absolute zero (0 degrees Kelvin or minus 273.15 degrees Celsius), can exist in such a state in perpetuity.[2] The international press reported a startling discovery of virus components, trapped in an ancient ice lake in Dry Valley, Antarctica. Halophiles, 650 million years old, trapped in crystals of salt are viable and could be cultured, or grown,

[2] Morowitz, HJ, *Beginnings of Cellular Life*, Yale University Press, 1992.

today.[3] Furthermore, Hoyle and Wickramasinghe contend that DNA can survive the environments within comets – unless they travel too near to the sun. If DNA-carrying comets were to pass through our atmosphere, the DNA would be able to withstand this form of 'flash heating upon atmospheric entry', as Hoyle and Wickramasinghe put it, during the high temperatures caused by atmospheric resistance. This incredible longevity of organisms, under immensely hostile conditions, adds considerable weight to the Hoyle–Wickramasinghe hypothesis. Furthermore, Hoyle and Wickramasinghe contend that only one microorganism in every 10^{20} (1,000 million-million million) with an origin in interstellar space needs to survive for their panspermia hypothesis to be valid – a very small margin of survival indeed.

Although not the first to propose a modern panspermia origin of life – the Swedish astronomer and Nobel Prize winner[4] Svante August Arrhenius (1859–1927) did so in 1908[5] – Hoyle and Wickramasinghe were the first to produce a modern hypothesis based on cometary structure.

The traditional terrestrial argument has far more proponents than its nonterrestrial counterpart. Yet even the concept of simple organic molecular compounds forming in a primeval soup, later evolving into complex bacteriological life forms, has been adapted to accept that at least some of the components for life were brought to Earth by cometary and asteroidal bombardment. Current theory states that Earth was severely pummelled and indeed shaped by millions of vast impactors perhaps 500–1,000 kilometres in size, shortly before (in geological terms) life evolved: approximately 4,000 million years ago. This sort of bombardment literally changed the planet, making it incredibly hot, the impactors delivering material that swelled Earth's mass and therefore its gravity. The core of the planet formed in this way, driven by the process of decaying radioactive materials.

Until 1986, when the European Space Agency's Giotto mission was able to travel to and probe Halley's Comet using spectrographic analysis (detecting wavelengths of energy specific to known minerals and materials), this terrestrial-origin theory stated that all of the

[3] Dombrowski, H, 'Bacteria from Paleozoic Salt Deposits', Annals, New York Academy of Sciences, vol. 108, pp. 453–60.
[4] Arrhenius won the Nobel Peace Prize for chemistry in 1903.
[5] Arrhenius, SA, Worlds in the Making, 1908.

vital ingredients for life – especially carbon, hydrogen, nitrogen and oxygen, which form complex molecules (polycyclic aromatic hydrocarbons or PAHs) grouped under the term 'organic' – originated in the molten core of our planet. After the Giotto mission, it was clear that Wickramasinghe's decade-old arguments (1974) about the organic composition of cometary dust tails were correct.[6]

The panspermia concept is seductive, but one must guard against the allure of fresh ideas. The terrestrial neo-Darwinian origin theory and the panspermia hypothesis are unproven, both potentially specious. Only time and scientific study will solve this argument, or offer another hypothesis to dismiss both as fallacious.

Such is the contention within the scientific community, however, that space-probe missions are being mounted that will bring the panspermia hypothesis to task; most notably NASA's Stardust mission (launched in 1999) will reach the comet Wild 2 in 2004. Stardust, as the name implies, will collect tens of thousands of dust particles from the comet's dust tail on a sticky gel pack, which will then be returned to Earth for analysis at the University of Kent.

Whatever the future holds, creationists will be heartened by the possibility that the formation of PAHs and other materials, perhaps even bacteria and viruses, found in these comets, may still one day be proven to be the handiwork of the omnipresent Almighty.

Life is a mathematical chain, a twice-coiled conundrum, the double helix of DNA, analogous to a vast computer code decreeing the nature of structure, the pattern of essence. Sound, light, even the electrical impulse of a heartbeat, operate in a quantifiable way, and so perhaps does evolution. If so, then mathematical models constructed to give us an understanding of evolution, and the role of catastrophe in that process, might enable us to speculate on how humankind's destiny has been shaped by impact.

Just such a model has been recently constructed by Satoshi Chiba of the Institute of Biology and Earth Science, Shizula University, Japan. The Chiba model[7] was designed to develop an understanding of the physical change processes that occur in population groups

[6] Wickramasinghe, C, 'Polyformaldehyde Polymers in Interstellar Space', *Nature*, vol. 252, no. 5483, 6 December 1974.
[7] Chiba, S, 'A Mathematical Model for Long-term Patterns of Evolution: Effects of Environmental Stability and Instability on Macro-evolutionary Patterns and Mass Extinctions', *Paleobiology*, 24 (3), 1998, pp. 336–348.

faced with environmental disruption in their habitats. Chiba looked at the likelihood of extinction under various scenarios of environmental fluctuation and change. He assumed that population growth was an ongoing process which is perturbed by random disaster.

Bearing in mind his caveat that the model is not designed to 'replicate' nature, but to identify the importance of environmental instability and the role it plays in evolutionary patterns, his findings were conclusive. It is apparent that creatures with small bodies, 'high fecundity and simple forms evolve in unstable environments'. Creatures that have developed a high degree of adaptation for their own, stable, habitat over a long period of time are most prone to extinction.

As Chiba puts it, 'The hypothesis that mass extinctions have occurred during periods of environmental stability is consistent with the fossil data. Adaption to a stable environment makes the taxa (population) vulnerable to environmental change.'

The more stable the environment, the greater the risk of extinction when a significant flux in the prevailing environmental conditions occurs – such as in an impact. It is clear that mass extinction, such as that seen at the K/T boundary, is most likely to result from a high degree of evolutionary adaptation to a very highly stable habitat.

If one accepts for a moment that Chiba's model is a workable interpretation of reality and that a vast comet/asteroid bombardment did occur (and there is little opposition within the scientific community to that theory) and continued throughout geological time – albeit in a somewhat less violent form – then that same bombardment process may, billions of years later, have helped shape our human history. Chiba's model may help to solve a conundrum of supremacy: how did *Homo sapiens sapiens* become the dominant hominid? It is possible that continual habitat change, thanks to impact, may have enabled *Homo sapiens sapiens* to supplant rival hominid groups.

At one point in human history there were, almost certainly, two types of human being: *Homo sapiens neanderthalensis* (Neanderthal) and *Homo sapiens sapiens* (the modern human). They were extremely similar, lived alongside each other, possibly in rival communities, yet one species did not survive, while the other did. Why?

Neanderthal man was the dominant hominid in West Asia and Europe between approximately 150,000 and 40,000 years ago. This 110-thousand-year rule (longer than modern humanity's dominion over Earth) ended when the higher evolutionary *Homo sapiens*

sapiens replaced the Neanderthal, approximately 40,000 years ago. *Homo sapiens sapiens* may have developed in another area – probably Africa – having moved away from the main hub of Neanderthal society, evolving separately in a new environment.

Piere Vermeersch, an anthropologist from the University of Leuven, Belgium, discovered a badly decomposed skeleton, thought to be 50,000 years old,[8] with the aid of his archaeological team. This necrotic child may have been buried in southern Egypt, during one of the most crucial periods in human history: the uprising of *Homo sapiens sapiens* and the overthrow of the Neanderthal.

Two rival theories exist about the origin of humankind. Some archaeologists and anthropologists believe that we spontaneously evolved in diverse places, achieving similar evolutionary adaptations despite inhabiting vastly different environs. This is possible, though unlikely. More logical is the much vaunted 'Out-of-Africa' theory, which states that *Homo sapiens sapiens* developed in Africa and migrated, possibly because of environmental perturbations, to the rest of the world. The location of the Vermeersch child is on one of the routes thought most likely to have been a migratory path from Africa to Europe: Taramsa Hill, southeast of Cairo. Although arguments will persist about the true nature of the child – *Homo sapiens sapiens* or Neanderthal – for some time, those who have actually seen the body, including Vermeersch, are convinced that its anatomical composition is identical to that of a modern human. This would be by far the oldest modern hominid found on the suggested migratory path from Africa.

If Vermeersch's child is indeed some poor early-modern human who died early in life, then that human will lend tremendous weight to the African-origin theory. So, why did humankind leave Africa, its birthplace, the cradle of humanity?

As we see in our modern world, when an intelligent predatory animal such as the lion is crowded out of its natural habitat, thanks to either man's development of the terrain or an environmental catastrophe – such as droughts caused by El Niño[9] – the lion begins to hunt

[8] Copley, J, 'Nearly out of Africa', *New Scientist*, 2155, 10 October 1998.
[9] A series of climatic changes occurring in the Equatorial Pacific region, originally applied to a warm current affecting the ocean off northern Peru and Ecuador. This warm current is now recognised to be part of the wide phenomenon, which has an often unpredictable effect on the world's climate.

for new sources of food: prey that would not ordinarily be included in its diet. It will even attack humans on occasion. Yet the beasts are not the only form of animal to fight for their territory and existence: humans have also fought, this century, when change beckoned. Although this change may have had political or imperial origins, people's homes, their nationality and what might therefore be defined as their identity, were threatened by annexation or war. The two World Wars were fought for territory and the suppression of tyranny.

Race violence has also been common this century: the civil-rights movement in 1960s America and the struggle against apartheid in South Africa until the 1990s are good examples. In Britain, race is an issue seldom far from the political agenda. The nation's police forces, for example, are constantly accused of bigotry and discriminatory practices – which have, in some cases, been acknowledged as true. What has this to do with our ancient ancestors and the destruction of our cousins the Neanderthals?

If we react as the lion would when our habitat is threatened or even when a new culture joins with ours, how would Neanderthal and *Homo sapiens sapiens* have reacted when thrown together?

Clearly *Homo sapiens sapiens* left Africa for a reason. If an environment is womblike, a garden of plenty, there would never be a reason to adapt, to pursue new pastures. Yet if that environment were suddenly and irretrievably changed or destroyed, an Eden no more, then survival would depend on locating new surroundings. If those new surroundings already harboured intelligent predators, but the area was bountiful, an agreement might be reached whereby both groups could coexist, then all would be well. This may indeed have occurred when *Homo sapiens sapiens* left the cradle of Africa, living alongside European and Asian Neanderthals. Then the disaster may have returned, perhaps thousands of years later, creating another dilemma: suddenly resources in this new pasture were scarce. Clearly only one race could survive. A race war – a pogrom – would have ensued.

No one can be sure of the story of our origins. Nor can we be sure of the regional catastrophe or environmental change that pushed *Homo sapiens sapiens* from Africa. What we do know is that this migration and ascension of humanity occurred during the last great glaciation: the Ice Age.

We should have some sympathy for our distant ancestors, as currently our world's climate is changing in the most perplexing manner, even from year to year. As ours changes, so too did theirs, though in a much more dramatic way.

Great sheets of ice had crept from the poles, encroaching on hitherto temperate and fertile lands, turning forest into ice desert from around 1,000,000 BC to 100,000 BC. It had happened before, but during the last ice age, humans were ascendant. Why? Perhaps because the hominid is very adaptable, coping with change: some of its kind may have died but those with an evolutionary edge survived. The ability to hunt increasingly scarce food and to develop fire and other technologies such as warm clothing were the keys that opened the doorway to survival. This great freeze could have been caused by a number of factors: declining solar activity, changes in ocean currents, or perhaps, as Victor Clube and his colleague DJ Asher, from the Communications Research Laboratory, Japan, have suggested,[10] the world may have been driven into an extended period of cooling (approximately a million years) thanks to cometary debris. Not an impact with Earth, but a protracted series of bombardments that delivered such quantities of ice and dust into our atmosphere as to effectively trigger an ice-box effect – akin to an impact winter, but caused by minute motes of submicron dust over hundreds or thousands of years.

In 1998 Clube and Asher went further and proposed that the cooling events have continued to this day and that they may be caused by the Beta-Taurid stream of meteoroids (also known as the Beta-Taurid Complex), which were in turn created by the giant comet Encke in a primitive and much larger state: proto-Encke.[11]

Clube and Asher believe this progenitor of catastrophe, trapped within our solar system, would account for the ongoing enhancements to the 'stock' of meteorites and comet fragments, of the Tunguska type, that pummel Earth. Further, they believe that proto-Encke's orbit has resulted in catastrophic increases in eras of Earth bombardment, which last for approximately 200–400 years in every 2,500–5,000-year period. So in every 5,000-year period a bombardment will begin and will then continue for up to 400 years. In this

[10] Asher, DJ, and Clube, SVM, 'An Extraterrestrial Influence During the Current Glacial-interglacial', *QJR Astron. Soc.*, Vol. 34, No. 481, 1993.
[11] Asher, DJ, and Clube, SVM, 'Towards a Dynamical History of Proto-Encke', *Celestial Mechanics and Dynamical Astronomy*, Vol. 69, 1998, pp. 149–70.

time the tremendous quantities of meteoroidal dust and debris that filter into the atmosphere build up to form a reflective shield, throwing the sun's energy back into space, changing, in effect, Earth's reflective abilities – known as its albedo.

Think of light passing normally through a pane of glass in a kitchen window, warming a room and nourishing a plant, by encouraging photosynthesis, on the other side of the window – encouraging life. However, were the owner to fall ill and the window was never cleaned, it would become dusty to the point where it eventually blocked the penetration of sunlight, resulting in a cooler room and a light-starved – dying – plant. So may proto-Encke's dust debris affect Earth.

The reverse is also true: where the dust thins and clears as Earth enters a protracted period (2,500–5,000 years) without intensive dust loading, global warming may result. These cycles were derived from the believed orbital activities of proto-Encke. The orbital predictions tally with the emergence of *Homo sapiens sapiens* and his cousin Neanderthal during the Upper Pleistocene period of geological history (100,000–10,000 years ago), therefore directly influencing Earth's climate during the last ice age. It is likely, if Earth's atmosphere was being heavily bombarded with submicron dust at this time, that small meteoroids and Tunguska-type objects actually impacted with the planet's surface or exploded (as in the Tunguska case) close to the surface. This form of repeated regional catastrophe could well account for the destruction of *Homo sapiens sapiens'* Eden and be the causal factor in his quest to find pastures new in Europe and Asia. Conflict with Neanderthal is pure speculation: it is possible that Neanderthal's gene pool was severely limited, perhaps amounting to only a few thousand individuals, though this seems unlikely when considering the geographic spread of Neanderthal's population – west across Europe to Britain and east into Asia. It seems not unlikely that adaptable *Homo sapiens sapiens*, having survived continual hardship in Africa, was more able to survive in harsh conditions that may well have occurred in Neanderthal territory. I think it possible that xenophobia, perhaps at first submerged, became a true motivating factor when food and good hunting territory became too scarce for competing intelligent hominids. The sturdy *Homo sapiens sapiens*, already hardened by his ancestral experiences in Africa, won.

My impact argument as a causal nexus for ongoing environmental change and regional catastrophe, thereby affecting humankind's pre-history, is enhanced by the high improbability of other, very familiar, mechanisms of catastrophe playing a part. Hurricanes, for instance, are a tropical phenomenon, and are extremely unlikely to occur during a period of glaciation/ice age. Tropical cyclones or hurricanes are generated over warm oceans by the heat released during the condensation of water vapour. Temperature is critical: below 27 degrees Celsius hurricanes weaken or fail to develop over the oceans.

Volcanic activity was infrequent in the late Pleistocene, possibly because of the additional surface pressure exerted by the glaciers. As the glaciers retreated and sea levels rose, so volcanism increased – as if the loss of this great weight allowed the latent volcanic pressures to break through.

Earth tremors may well have been ever present, but by far the most likely explanation for the sacking of early humanity's African Eden, if one accepts Clube and Asher's hypothesis, is that of bombardment from space and regionally devastating impacts of varying energy yields, from the very small to Tunguska-sized events – thanks to the return of proto-comet-Encke.

Our sympathies for our distant relatives may have modern-day echoes, more pertinent even than the El Niño example I gave. Today we, the First World, and the developing nations are gripped by the frightening concept of global warming. This is a measurable phenomenon, a very real event. In Britain, the land of drizzle and mist, our summers have become eccentric to the point of absurdity, our winters vacillate between intense cold and snow and springlike conditions. Britain's Benfield Greig Hazard Research Centre, a department of University College, London, has produced a report that states that there has been a 0.4 per cent rise in global temperatures since 1970.[12] Dr Mark Saunders, author of the report, told me, 'Global temperatures are higher now than they have been at any time since 1400 AD. Global warming is undeniable. As carbon dioxide [CO_2] levels in the atmosphere have risen from around 280 parts per million in the pre-industrial period to 360 parts per million in 1997, it is evident that our activities have contributed to the climate change.'[13]

[12] Saunders, M, *Global Warming: The View in 1998*, Benfield Greig Hazard Research Centre, 1998, p. 9.
[13] Saunders, M, interview, 1998.

Saunders's evidence for atmospheric CO_2 levels in the preindustrial era originates in ice cores, similar to those taken during the GISP2 mission in Greenland, though in this case from the South Polar extremes. While the current global-warming phenomenon may indeed be linked with escalating human activity, it may also be connected to the periodicity of the Encke dust veil: currently Earth may be enjoying a respite from the dust debris and severe atmospheric loading of the kind that Clube and Asher suggest triggered the last ice age. Certainly, current historical chronology dictates there was no modern-type industrial activity prior to the so-called Industrial Revolution, which most economic historians believe began around 1740–80, when the economy of Britain (the first industrial nation) began to grow by more than 2 per cent per annum.

Clearly, according to historical tradition, the last warming period had little to do with humankind, unless there was some spectacular lost civilisation pumping CO_2 into the atmosphere around 10,000 BC. It is surely not impossible that a civilisation emerged at an earlier date and was lost, due to some natural catastrophe, but it remains unlikely. It seems possible, then, that humankind's current contribution to global warming, heinous though it is, may not be the main cause of Earth's rising temperature.

Clube believes that, in time, it will be proven that our climate, environment and, in many ways, daily existence are directly influenced by atmospheric loading, with impacting bodies and dust, from space: 'It is the flux of these bodies,' Clube states, 'that is critical to the everyday existence of the human race.'[14]

Encke's icy grip may have left Earth for the moment, but it is most likely to return, and its deadly offspring, the Beta-Taurid meteoroid stream, continues to threaten Earth every June and November.

Indeed, the Beta-Taurid complex and other space flotsam – there are sixteen major meteor streams that shower the northern hemisphere alone every year, though the Beta-Taurid complex may be a million times larger than an average meteoroid stream – may be responsible for far more of our history than one might first expect.

The Holocene (current time) period began at the end of the Upper Pleistocene and dendrochronologists have produced evidence that lends weight to the ongoing space-origination and environmental-

[14] Clube, SVM, interview, 1998.

perturbation argument. Mike Bailey, of Queen's University, Belfast, has noted that widespread events of surprising brevity occurred in the recent past which had no apparent terrestrial explanation,[15] particularly the 1740 event. This event was catastrophic and followed a terrible winter in 1739, lasting a full seven weeks, resulting in mass famine and the death of approximately 300,000 people. Ireland's tree-ring data is conclusive and when Bailey compared his data with that of other researchers[16] it became apparent that the experience in Ireland was echoed across Northern Europe in 1739–41. Searching for a cause, Bailey has come to regard the impact phenomenon – most particularly the loading of the atmosphere with cometary/meteoroidal dust – as the most probable explanation.

There are far more instances than the 1740 event that seem likely to have been triggered by impacts or cometary dust-loading instances. One of the most remarkable not only appears to have influenced climate (as evidenced in tree growth in Ireland) but was also recorded in history by one of Rome's leading historians and essayists: Magnus Aurelius Cassiodorus. The date was about AD 534–6, and Europe had begun to face a severe downturn in climate with resultant famines. The weakened population – no doubt suffering from depressed immune systems thanks to malnutrition – was badly hit (circa AD 542) by plague.[17]

Cassiodorus wrote of a strange phenomenon that was causing considerable consternation among his people and one that may offer a remarkable insight into the origin of the 534–42 problems. He wrote:

> ... who will not be disturbed, and deeply curious ... if something mysterious and unusual seems to be coming on us from the stars? For, as there is a certain security in watching the seasons run on in their

[15] Baillie, Dr Michael, *Dendrochronology Provides an Independent Background for Studies of the Human Past*, pp. 105–112.

[16] Drake, M, *The Irish Demographic Crisis of 1740–41*, Historical Studies VI, Keegan Paul, 1968, pp. 101–24, and Post, J, *The Last Great Subsistence Crisis in the Western World*, Johns Hopkins University Press, 1977.

[17] Baillie, M, 'Marking in Marker Dates: Towards an Archaeology with Historical Precision', *World Archeology*, Vol. 23, 1991, pp. 233–43.

succession, so we are filled with deep curiosity, when we see that such things are changing.[18]

Cassiodorus's questioning of apparent climate change sounds a familiar note. His comments would not be out of place in the letters column of a modern newspaper: even now we are preoccupied with the rapid fluctuations in our climate. While our meteorologists struggle to understand current global-warming trends, Magnus Aurelius's suggested origin for the perturbations was obvious and alarming:

> How strange it is, I ask you, to see the principal star [the sun], and not its usual brightness; to gaze on the moon . . . shorn of its natural splendour? All of us are still observing . . . a blue-coloured sun . . . we marvel at bodies that cast no midday shadow . . . and this has not happened in the momentary loss of an eclipse, but has been going on equally through almost the entire year . . . whence can we hope for mild weather, when the months that once ripened the crops have been deadly sick under the northern blasts? For what will give fertility, if the soil does not grow warm in summer?

What is one to make of such a striking account? A sun turned blue, no shadows cast at midday, terrible gales of icy wind and absence of the sun in summer? SJ Barnish, the translator of Cassiodorus's *Variae*, dates the writing of this passage to the spring of AD 534. If Barnish's date is correct, Cassiodorus's comments about the terribly cold climate and lack of clear sunlight for almost a year, and the fact that the soil was unable to grow crops in summer, tallies with the environmental downturn recorded in Bailey's dendrochronological data. The visual phenomenon Cassiodorus described, combined with the general haze and cooling, is highly indicative of an impact. The fact that the climate downturn was so severe that it had lasted for almost a year by the time Cassiodorus wrote his account – and according to dendrochronology went on to trouble plants and humans alike for nearly two years – suggests a tremendous perturbation of our planet's environment.

[18] Barnish, SJB, *Variae of Magnus Aurelius Cassiodorus*, Liverpool University Press, 1992, pp. 179–81. This translation is highly accessible, engrossing and recommended. Barnish has returned Cassiodorus to life with the efficacy of his work.

It is known from other records that the whole northern hemisphere was affected. While it is possible that a vast eruption of volcanic dust was responsible, the absence of an acid layer[19] or 'volcanic-fingerprint' in ice cores suggests that there may have been another cause: atmospheric loading from space, an impact. As more science is done, it will become evident which phenomenon is genuinely to blame, or whether such extraordinary events were as a result of more than one influence.

The notion that cometary debris has influenced our very evolution and our history, and that even today it may be doing so, is a frightening one. It would imply that what we once thought of as a certainty, the passing of spring into summer, then autumn and winter, a clearly defined seasonal system, might be only a temporary (in historical terms) lull in an ongoing process that operates on a timescale far in excess of the human life span, but which can result in abrupt changes, even from one year to the next. As we will see in Chapter 3, the ages of ice have been broken by ages of water. Few nations on Earth are devoid of a mythological flood, akin to the biblical episode, and evidence is extremely strong that these 'inundation events' have origins in impact.

If we are to try to understand what the future might hold, whether catastrophe beckons and the nature of its threat, then history and mythology must have a role to play: what is mythology but oral tradition, a primitive method of recording history? Humankind has long recognised the value of learning from past mistakes and bearing incidents of terrible disaster in mind when planning for the future. Prosaic mythology – such as great Titans clashing in heavenly conflict – often arose as an interpretation of real events, seen through the eyes of humans with only religious codes and superstition to offer them an understanding of the spectacles they witnessed. They are rationalisations from socioreligious cultures devoid of any other 'science'. Objects and events were often personified into anthropomorphic heroes, so as to allow for a greater understanding and ease of remembering of the events. Just such an instance appears to be the Phaethon myth.

One of the most outstanding interpretations of this myth is told by the Latin scholar Publius Ovidius Naso (Ovid). Born in 43 BC (a few months after the scandal of Julius Caesar's assassination rocked

---

[19] Baillie, M, *The Holocene*, 4, 2, 1994, pp. 212–17.

Rome), Ovid became one of Rome's foremost Latin scholars and poets, writing the *Metamorphoses* in AD 8. The *Metamorphoses* includes a brilliant description of Phaethon, the immature male offspring of the sun god Helios, who begged his father for the use of his chariot in order to fulfil a rite of passage to maturity. The tale tells us that the boy, weak and inexperienced, loses control of the great light-chariot and its winged horses, and how the chariot careers across the heavens only to impact with Earth, setting land and forest afire as it does so. This is a tale of a small sun or fireball impacting with the Earth:

The earth caught fire, starting with the highest parts. With all its moisture dried up, it split and cracked in gaping fissures. The meadows turned ashen grey; trees, leaves and all were consumed in a general blaze, and the withered crops provided fuel for their own destruction . . . Great cities perished, their walls burnt to the ground, and whole nations with all their different communities were reduced to ashes . . . Neptune tried to raise his arms . . . the fiery air was too much for him . . . the goodly earth goddess found herself entirely girdled by waters . . . she made everything tremble . . . then sank down . . . lower than she used to be . . . The omnipotent father sent forth a thunderclap . . . against the charioteer . . . he dashed him from his car . . . The fragments of the car were scattered far and wide . . . Phaethon, with flames searing his glowing locks . . . went hurtling down through the air, leaving a long trail behind . . . one day passed without the appearance of the sun: the burning fires gave light so that the disaster served some useful purpose.[20]

How could such a mesmerising tale be created but as a way of remembering the impact of a meteorite or asteroid, as an oral tradition? This striking text is but one of many that, while not forming part of written history as we understand it, did form part of history as the Ancient Greeks understood the word. Theirs was a great tradition of analogy and metaphor, and this metaphor seems to have ably recorded a major impact event somewhere in their distant history. The first Greek-speaking peoples established the Mycenaean civilisation, in Greece, during the Aegean Bronze Age, circa 1400 BC. They ruled the area for approximately 200 years, a fabulous dynasty

[20] Innes, MM (translator), *The Metamorphoses of Ovid*, Book II, Penguin Classics, 1955, pp. 60–2.

of advanced learning with great artistic tradition and the technology of writing. Indeed the technology of the written word died
with the Mycenaean people. Lost for nearly half a millennium,
writing was not rediscovered until the next Grecian epoch, following
a great deal of study of Phoenician texts.

No one understands the downfall and end of the Mycenaean
society and cities, but this great civilisation vanished. The decline
began in the thirteenth century BC and its degeneration triggered a
great dark age in one of the world's first centres of civilisation.

Can it be that the Phaethon myth tells us of that great civilisation's
destruction? Homer's *Iliad* reminds us that the Mycenaeans were a
warlike people, their sacking of Troy an immortal testament to their
antipathy. Perhaps, then, the Mycenaeans destroyed themselves.
Maybe some catastrophe resulted in a battle for scarce food and
fertile land resources: a civil war? But their sorrows are a long time
lost, and it is likely that we shall never know the truth of it. The
Phaethon myth stands, however, as a gripping account of some catastrophic fireball, destructive in a way that modern science can
understand, personified for eternity as the unfortunate offspring of
the day-bringing Sun himself.

Our knowledge of the Greek myths and pantheon of gods stems,
in part, from the work of Homer, the first great poet of Ancient
Greece. Legend has it that he lived at the time of the Trojan War, the
twelfth century BC, but modern scholars believe he lived (if he existed
at all) in the eighth century, born in the region of Ionia. If ancient
tradition is correct, however, the destruction of the Mycenaean civilisation would have taken place only one lifetime before Homer
committed the mythology of Greece to the written word. Homer has
long been acknowledged as a writer prone to symbolism; the Phaethon myth is extremely symbolic and might have been established in
something like its current form by Homer. Much of his writing is
thought to have been lost, but his work was read and reinterpreted
by scribes who came after him. Perhaps, therefore, if he heard the
tales of the Phaethon fireball, from relatives or others who had seen
it, he established the nature of its symbolic interpretation. If so, his
may be one of the earliest attempts to rationalise a catastrophic
impact.

It is possible that another legendary Greek scholar, Plato (428–
c. 347 BC), gave us insight into this catastrophe – without consciously
realising that he was doing so – in his dialogues *Timaeus* and *Critias*.

Many have suggested the Mycenaeans to be the true people of the Atlantis legend. I say this with caution as much hyperbole has been created surrounding this passage in Plato's work. This ongoing fascination with the Atlantis myth is astonishing, especially when one consults the original text, which surprises with its brevity. Plato's Atlantis story – recounted in the text by Critias, then said to be a ninety-year-old man, to Socrates – is also striking in its appropriateness to the destruction of the first Greek (Mycenaean) civilisation and the Phaethon myth.

The work describes a great warlike power, very like the Mycenaeans, who advanced from their territorial base in the Atlantic to besiege the cities of Europe and Asia. Their island, a vast nation, was said to be as large as Asia and Libya combined. If this sizing is anywhere near correct this island state had an area of approximately 44.46 million square kilometres.[21] Plato described a highly navigable Atlantic with islands dotted across the sea facilitating crossing between the continents (possibly Eurasia and the Americas). This warlike race that Plato said inhabited a place called Atlantis had tried to exercise imperial might, seeking to annex other nations. A war ensued and the people of Athens succeeded, according to Critias's tale, in repelling the Atlanteans. Later, the victorious Athenians and the people of Atlantis were forfeit – killed by an almighty catastrophe.

> . . . there were earthquakes and floods of extraordinary violence, and in a single dreadful day and night all . . . men were swallowed up by the earth, and the island of Atlantis was similarly swallowed up by the sea and vanished.[22]

This fascinating account has many similarities, though devoid of the Homeric symbolic style, to that of the Phaethon myth, in particular Plato's sequence referring to the damage to the land and perturbation of the waters as compared with Ovid's Earth goddess eventually sinking lower than she had originally been.

Plato's Atlantis myth may indeed be a retelling of the Mycenaean collapse, perhaps presenting a similar catastrophe to the Phaethon

---

[21] *Collins Pocket Atlas*, 1969, pp. 7 and 9. Libya has an area of approx. 1.76 million sq. km, while Asia covers an area of 42.7 million sq. km.
[22] Lee, D (translator), *Plato's Timaeus and Critias*, Penguin Classics, 1977, p. 38.

myth, but seen from a different perspective. It is possible that no connection exists, but the destructive catastrophe, ancient timing and descriptions of the Phaethon myth and the Atlantis story all offer intriguing possibilities for the destruction of the Mycenaean society. Were these phenomena related and caused by catastrophic impact? Perhaps we will know in time. The invention of new archaeological techniques, or some new discovery – the ruins of Troy itself were not discovered until 1870, by the German archaeologist Heinrich Schliemann (1822–90) – may yet unearth the truth. Whatever the outcome, the possibility that these ancient words are records of a catastrophe with celestial origin remains: this long-defunct epoch of human history may have been literally obliterated by impact. Even though this material offers circumstantial evidence of historical incident it remains strongly indicative of cause and effect and is, I feel sure, worthy of further cross-discipline study.

We might find a reminder of the fragility of our own long-lived civilisation in the millennia-old pages of Ovid's *Metamorphoses* and Plato's *Timeas* and *Critias*. This history ably demonstrates how, in epochal cycles, complacent civilisations have vanished, to become little more than dust on the winds of time. The problem is one of the structure of society: even primitive tribal societies demonstrate degrees of specialisation. Where a society becomes specialised, with individuals honing their talents in a particular field of enterprise, knowledge becomes fragile. We are all specialists in our own way, whether bank clerks or candle makers.

Few of us, save the candle maker, could make a candle from scratch, let alone survive as our animal relatives do. At first glance, making a candle appears a simple task – it might be one that we would need to understand if an impact destroyed our cities or our electricity supply.

What would one make a candle from? How to find and purify the wax, obtain the fibre strands for the wick, let alone twist and create the wick? Then how does one melt the wax and what would one cast it in? Indeed, how should one make a mould, and from which materials? Come to that, without electricity or gas supplies, how would one generate fire? In time there might be few matches. That is how fragile knowledge is: we might be reduced to developing the fire-making skills first employed by our ancestors during their apparent cohabitation with our cousins the Neanderthals.

Impact is a dangerous threat, one that has consequences far in excess of environmental destruction and loss of life. Our sum of knowledge, so vast that only a specialised society can wield it with efficacy, could be lost in an instant; only fragments would remain. Fragments may not be enough when we're challenged by starving packs of dogs, or faced with dysentery or cholera due to insanitary water.

Legends of Eskimo and Norse migrations in Greenland, in an age of climatic turmoil, have persisted since the fifteenth century. Evidence gleaned from ice cores obtained during the recent second Greenland Ice Sheet Project (GISP2), dendrochronological data, solar irradiance (solar activity) estimates and written history have proved that Earth was gripped by a so-called 'Little Ice Age'. This was not a short-lived perturbation, banished to some remote time: the Little Ice Age is thought to have stretched across nearly 500 years, from 1400 (or possibly 1420) to the 1890s. This is a phenomenon that some of our grandparents would remember: terribly harsh winters, cool summers and overall an era of extremes.

The Vikings and Thule Eskimos who inhabited Greenland in the fifteenth century had both migrated to this land, the Norse from the east and the Eskimo from the west. Both were colonisers, but the two had totally different socioeconomic structures: the Norse culture was one of farming and husbandry, the Eskimo (Inuit) one of the hunter-gatherer. The GISP2 project produced soil evidence that revealed the emergence of Arctic insects on Viking farms, replacing the insects associated with milder and more fertile climes. The Vikings inevitably moved ever westward in search of warmer places to live. The highly adapted Inuit Eskimo was able to hunt at sea, on land or even on the ice itself. Theirs was a way of life well suited to the new climate of this Little Ice Age. Their society, already adapted to a cold environment, was able to thrive. The Norse society, also highly adapted, but to a warmer, more fertile climate, had to migrate to survive.

This relatively recent migration westward, due to an environmental downturn – possibly caused by a dust loading of the atmosphere like that described by Clube, due to impact or cometary debris – may mean that the Norseman followed the example of our early *Homo sapiens sapiens* ancestors from Africa. They each headed to new garden-pastures in which to subsist and ensure the continuance of their societies. While their journeys were separated by approximately

50,000–100,000 years, the motivation for leaving their homes may have been the same.

In the eighteenth century, the Hudson River in New York Harbor was known to freeze so hard that it was used as an ice-skating rink by the local people, just as the Muscovites enjoy the Volga in winter today. Since the beginning of the twentieth century, temperatures have been on an upward trend.

These connections through time show a pattern of influence, creating the possibility that impacts have ended human epochs at particular moments in Earth's history, while encouraging the evolution of a different era in another. As the studies of Gunnar Heinsohn and Michael Baillie demonstrate, impacts and climate perturbation due to loading of cometary dust may also have steered the evolution of society and belief patterns. This is strongly reinforced by the vivid descriptions of atmospheric phenomena in historical documents such as Cassiodorus's *Variae* and the more metaphorical Phaethon myth.

The Cretatious/Tertiary (K/T) impact (65 million years ago) destroyed the dinosaurs and another such impact may have wiped out early human civilisations along with their science, culture and art. We are only now beginning to suspect that civilisations have occupied the Earth for more than a mere 7,000 years. Weathering patterns on ancient temples and cultural similarities between ancient civilisations found across the globe, suggest the existence of developed trading nations more than 10,000 years ago. Only at the beginning of the third millennium of the current calendar has Western science begun to swallow its pride and accept that civilisation may not be a recent phenomenon.

Long-regurgitated myths have recently been revealed to be true: fragments suspected to be remnants of the Colossus of Rhodes and underwater structures – a sunken city (Mu?) – off the coast of Okinawa have been discovered. Pottery in Japan has been revealed to be at least 12,000 years old, and new archaeological evidence proves that 'man' first reached Australia 40,000 years ago. Is it too much to assume that there were civilisations in existence between 40,000 BC and 5,000 BC? If so, some catastrophe, whether political (war) or accidental (asteroid impact), wiped much of the evidence of their existence from the surface of Earth.

It seems likely that one or more collisions with heavenly bodies have steered the course of human history – perhaps creating whole separate epochs, defined by loss of knowledge and the rediscovery of science and society. Humans may have lived on landmasses other than those we know today. Legends of lost cities swallowed by oceans, such as Plato's Atlantis, may have their origin in ancient impacts. We live on only 28 per cent of the Earth – 72 per cent of Earth's landmass remains an aquatic mystery. Humankind has only recently, in the twentieth century, begun to learn about the mysteries of the deep – the first accurate map of the world's undersea floor was not developed until 1997. What archaeological evidence remains buried under the sediment of the oceans?

If a 15-kilometre object impacting with Earth could virtually eradicate aeons of dinosaur history, why could it not also erase mere tens of thousands of years of human achievement? Much as all trace of the dinosaurs, which roamed Earth for 400 million years, has been lost – bar a very few fossil remains – so too could early human civilisation. Even the science of palaeontology is relatively new – discovery of the existence of dinosaurs was made as recently as 1822, when Gideon Algernon Mantell (1790–1852) first discovered fossilised dinosaur teeth and bones in Sussex.

Evidence of earlier human epochs may be buried under tons of rock – thanks to impact ejecta – or miles of ice, in our polar regions, as a consequence of an ancient impact winter. More worrying still is the thought that this, the most modern of humanity's civilised epochs, could also be lost, little left but fragments of a society that could split the atom and engineer life.

Do we really want to lose all that we have gained in this period of abeyance from catastrophe?

THREE

AGES OF ICE AND WATER

Could there be a harsher world than one covered by permanent sheets of ice? I cannot imagine a more horrifying place. What more disturbing prospect could there be than living out one's life knowing that our children will never see the fields of green that we once enjoyed, nor will they delight in the bounty of a prosperous world bathed in education, industry, travel and, most importantly of all, warmth. A world gripped by an ice age, perhaps 100,000 years from beginning to end. A period so impossible to comprehend, so protracted, that legends and gods could rise and be forgotten. The recorded history of a long-lived civilisation might be lost in such a period; indeed a new sub-species of human could emerge. Such a world once existed, not on some fanciful far-off planet in a distant solar system but here, on this Earth, a land once frozen in the grip of the marching glacier.

This Pleistocene Ice Age was not a singular event. Our Earth's past is littered with periods of glaciation, so frequent in geological terms that geologists refer to our modern epoch in history as the 'current interglacial', implying that another age of ice awaits us. The last ice age lasted for 100,000 years, ending just 10,000 years ago as Earth entered the Holocene or modern period. It can be little coincidence that recorded history began a little over 8,000 years ago as humankind, freed of the strictures of a meagre existence on icy plains, was able to develop society and permanent settlements. Perhaps this recent emergence from the age of ice into one of societal and economic structures reflects the difficulties in existing, let alone thriving, in a world that is so cold that its seas actually retreat; falling in the Pleistocene by as much as 130 metres. If humankind has been able to develop writing, mathematics, technology, even trading and farming in less than one-tenth of the equivalent period of the last ice age itself, what might happen to our society if we were to fall prey to another epoch of ice?

Evidence of the ice incursions across our continents amount to till (boulder clay) or loess (dustlike clay and silt) deposits left by glacial movement, and 'sculpted landscapes' carved by glaciers as they edged through the land creating huge glaciated valleys and, later, rolling drumlins. It is unequivocal: the world has frozen and thawed on at least five occasions. Today we stand at another climatological crossroads: 10,000 years have passed since the end of the Pleistocene; the geological deadline for the next great age of ice is imminent. Could we avert it if we come to understand the trigger mechanism: impact?

A chain of impacts, over a period of several thousand years, could have precipitated the ice ages. If so, the impact mechanism may be more complex than previously suspected, not just creating impact winters, but also triggering whole epochs of ice lasting hundreds of thousands of years. The trigger mechanism is complex and I feel sure that no one event could be entirely responsible for an ice age. It is far more likely that a series of catastrophes must take place in order to throw our world into a protracted glaciated state as has happened on those five[1] previous occasions in Earth's history.[2] Victor Clube's hypothesis that the nucleus of the Beta-Taurid meteor stream (responsible for shooting stars seen every year in November skies) could have caused the last ice age (100,000 years ago), and could do so again, stands. Our world could be laid waste by the advance of glaciers, thrown into a new epoch of ice that could last for 4,000 generations.[3] This is no slow transition over millennia: geologists believe that ice ages develop in less than a century, decades, or perhaps just a few short years.

Other rationales do exist for the generation of ice ages; all are old and have evolved from the uniformitarian school. Nevertheless, two key arguments have significant value and may be closely involved with the periodicity of impacts. In 1930, Milutin Milankovitch

---

[1] Osborne, R, and Tarling, D, *Historical Atlas of the Earth*, Viking, 1995, p. 152.
[2] The Pleistocene glaciation began about 1.7 million years ago, although the period referred to as the last 'Ice Age', a period during which approximately 25–50 per cent of the Earth's continents was covered with ice, began approximately 100,000 years ago. The other four ages of ice began 290–300 million years BC, 420–30 million years BC, 660–70 million years BC and approximately 2,000 million years BC, respectively.
[3] Assuming a generation to be 25 years.

(1879–1958), a geophysicist of Yugoslav origin, theorised that the Earth's cold phases may be the result of cycles that our planet moves through in space. He proposed that the Earth's orbit and its degree of inclination towards the sun may combine with oscillations about the planet's axis. The net result of these three cycles would be a reduction or increase in the quantity of solar radiation hitting the planet. In periods of higher exposure to the sun's rays, temperatures would increase and vice versa. Milankovitch estimated that global temperatures would fluctuate in a range of 10 degrees Celsius (5 degrees plus or minus) about the norm – enough to ensure a significant cooling of the Earth's surface. He perceived the problem to be one of *perpetual* winter, as opposed to merely harsher winters and cool summers. Winters would be much the same as now, but they would never end: there would be no thaw and warming summer.

The Milankovitch theory stands as a likely contributor to the ice-age phenomenon. Its effect may be limited, however, to regulating the periods of climate change: an ice age every 10,000 years. It is entirely likely that Earth's Milankovitch oscillations will ensure colder periods which will then be intensified by impact and dust loading of the atmosphere. The result: an ice age. Other factors may play a role: volcanic loading of the atmosphere works in much the same way as cometary debris and impact dust, hurling great plumes of volcanic ash and soot into the stratosphere, where it can remain in this aerosol form for many months.

One such eruption in recent times took place in the Philippines at Mount Pinatubo on 12 June 1991. This spectacular volcanic outpouring lasted for four days and was presaged by numerous explosive outbursts of energy that threw 15–20 million tonnes of dust into the stratosphere, reaching altitudes of 20 kilometres. The volcano was easily the largest known terrestrial producer of soot and ash from a single event, and triggered a measurable cooling of our planet's average temperatures.

Models of the Pinatubo eruption have produced startling data, suggesting a cooling thanks to the reflective property of the ash and soot, of 0.4 to 0.6 degrees Celsius. Physical measurements taken in 1992 compare favourably to the model, revealing a cooling of between 0.3 and 0.5 degrees Celsius.[4] Remember, this is not a

[4] Saunders, MA, *Global Warming: The View in 1998*, Benfield Greig Hazard Research Centre, 1998, p. 9.

regional or countrywide phenomenon, the whole planet's average temperature dropped because of that one event. Fortunately, 0.5 degrees is too little to result in ongoing consequences, according to current models. Nevertheless, this startling statistic shows that even relatively familiar events can have direct consequences on our world as a whole.

Significantly, the largest hole in Earth's protective ozone layer – a region of the upper atmosphere filled with ozone gas, which absorbs much of the harmful ultraviolet light that forms part of the sun's output – was detected in 1992–3. If the ozone layer were to be irretrievably damaged, the sun's ultraviolet rays would cause skin burns and cancerous tumours on all human and animal life, and would make the production of food crops virtually impossible.

The implications are worrying: Pinatubo's explosive release of energy and dusting of the atmosphere probably damaged this vital protection and caused a small but significant downturn in global temperatures. If Pinatubo could have such inordinate effects it is easy to understand how an impact, or a series of subcritical impacts, could generate major climatic perturbations.

Further evidence of the cooling effects generated by soot and ash can be gleaned by studying weather-satellite data from 1996 and 1997.[5] The Indonesian fires, set by human beings in the country's rainforests to clear land for cultivation, raged out of control in 1997. The fires are part of an agricultural tradition, but the consequences of climate perturbation caused by the El Niño effect were raging fires without the corresponding rainy season expected in the autumn. The lack of rain meant the fires raged out of control, releasing vast amounts of soot and ash in aerosol form. The volume of material was so great that Malaysia was also shrouded in the fallout from the fires, resulting in almost toxic air supplies for both countries' inhabitants. Most interestingly of all, however, despite the fires in Indonesia, is that a comparison between the satellite charts for November 1996 and November 1997 reveals a startling cooling event in the footprint of the fires' dust cloud, which stretched westward into the Indian Ocean. The Indonesian fires appear to have resulted in regional cooling of 5–7 degrees in November 1997.

There have certainly been numerous cold events in Earth's past –

[5] Satellite anomaly charts prepared by the USA's National Oceanic and Atmospheric Administration.

similar to these measurable cooling events this century but of much higher magnitude – which have pushed Earth from a temperate interglacial period, analogous to our own climate, into a period of glacial activity and intense cold. Heinrich Events,[6] as they are known, have occurred most frequently in the geological record. A key event occurred in the last interglacial period – circa 100,000–135,000 BC – and should serve as a warning, buried in the sediments of prehistory waiting for us to understand its import. The last interglacial was stunningly similar to our period, warm and temperate, though it is possible that sea levels were higher than today. At around 120,000 BC the climate had a severe downturn, resulting in great temperature drops and glacial activity over an area stretching from Ireland to China and as far south as Antarctica, though the primary area of perturbation was in the North Atlantic.

It is evident that lower average levels of salt in the sea, as the result of an introduction of vast quantities of fresh water, caused a fall in the creation of cold, dense water (water with high salinity has a higher boiling point than fresh water). This denser water ordinarily falls to deeper levels, creating an inrush of warmer surface water from equatorial regions: the result is warm surface seas and therefore warm coastlines. Because salinity levels dropped, the coastlines bordering the North Atlantic were no longer warmed by this maritime effect and their temperatures dropped rapidly. Three American environmental scientists, Adams, Maslin and Thomas, speculated in a recent report[7] that the increase in fresh-water levels might have been as a result of increased rainfall, due to the Milankovitch oscillations. Although raised levels of precipitation may have occurred, it seems not unlikely that cometary bombardment may have been the true source of fresh water, by introducing frozen water from space into the Earth system.

The Milankovitch oscillations might result in Earth's passing through dust rings – extraterrestrial matter accreted into a complex series of discs orbiting at various distances round the sun. One model of the solar system's dust rings, prepared by the American Carnegie

[6] Heinrich, H, 'Origin and Consequences of Cyclic Ice Rafting in the Northeast Atlantic Ocean During the Past 130,000 Years', *Quaternary Research*, Vol. 29, 1988, pp. 142–52.
[7] Adams, J, Maslin, M, Thomas, E, 'Sudden Climate Transitions During the Quaternary', http://www.esd.ornl.gov/projects/gen/transit.html.

Institution of Washington, ably demonstrates that dust debris could result in the 100,000-year periodicity of ice ages, as implied by the geological record and the Milankovitch theory. The Carnegie model[8] may yet prove a linchpin in achieving a new synthesis between the schools of thought that support volcanism, plate tectonics, the Milankovitch cycles and the impact mechanism as causalities for the ongoing ice ages seen over the last 1.2 million years.

It is possible that the impact phenomenon may actually help to stabilise our climate: too little dust in the atmosphere may result in global warming; too much and global cooling occurs. Similarly, the type of impact ejecta that is hurled into the atmosphere might determine whether an impact could cause an increase or decrease in temperature. Water might contribute to the 'greenhouse effect', which depends upon certain gases, such as carbon dioxide (CO_2), and liquids, such as water (H_2O), to retain the sun's energy. This effect is crucial to maintain the Earth's current temperate climate, elevating the planet's surface temperature to an average of 15 degrees rather than minus 18 degrees Celsius: 33 degrees warmer in total. Yet, if these gases and liquids are present in too great an abundance, the planet begins to overheat: as has been the case in the late 1990s when greenhouse gases have contributed to the phenomenon generally referred to as global warming. Indeed, recent analyses of global temperatures have produced a quantifiable increase in temperature thanks to the release by humankind of CO_2, which is largely due to our enduring fascination with the combustion of fossil fuels.

Dr Mark Saunders of the Benfield Greig Hazard Research Centre believes that global temperatures will increase by an additional 2–3 degrees Celsius by the end of the twenty-first century, with the probability of even greater temperature increases over land. He also estimates that sea levels will rise by an additional 15–85 centimetres and rainfall by another 5–25 per cent (especially over northern temperate regions such as Britain and Europe) by 2100.

Certainly, over the last hundred years global temperatures have increased by 0.06 of a degree Celsius, resulting in a significant retreat of glaciers and a rise in sea levels of 10 centimetres. Similarly, rainfall

---

[8] 'Cosmic Dust May Trigger Ice Ages', *San Francisco Examiner*, 10 September 1998.

has increased by 10 per cent in many high- and mid-latitude areas of the globe.[9]

A number of scientists now support the hypothesis that smaller impacts – of a size ranging between 0.1 and 1 kilometre in diameter (henceforth called subcritical impacts) – could actually cause significant seasonal change, thereby triggering glaciation or, more perplexingly, global warming. Professor Claudio Vita-Finzi of University College, London, defined the mechanism during a recent lecture:

> The effects of an impact will be determined by the location of the impact site. If a bolide [impactor] were to collide with land ice the effect would be worse in the northern hemisphere as opposed to the southern, because of the thickness of the ice . . . particulate ejecta from icy targets will contribute in winter months to the impact-winter effect that is associated with large collisions . . . Water vapour derived from thawed permafrost or melted sea ice could enhance the atmospheric greenhouse effect in summer months . . .[10]

The Ice Age demanded hardy individuals, well used to fending for themselves, experts in killing, skinning and preparing food – survivalists adept at living in a world alien to us. Perhaps the modern Inuit (Eskimo), or his grandfather at least, could relate to the *Homo sapiens sapiens* of the Pleistocene. Their lives were similar; perhaps the Inuit might look upon other races of modern humans in much the same way *Homo sapiens sapiens* looked upon the Neanderthal: just another group of people trying to survive as they see fit. The Inuit depend upon caribou, seals and even fish taken from ice-covered seas. Although the modern Inuit of the Siberian tundra and Alaska now seek to pursue Western lifestyles, their Westernisation is barely a generation old; still they carve bone and teach their young to hunt. Their frozen world is not then too dissimilar to that of our forebears: the glacial and postglacial human.

Imagine our forebears' world, defunct for nearly 10,000 years: intense reflected light, endless cold, the warmth of fire an essential

---

[9] Saunders, MA, *Global Warming: The View in 1998*, Benfield Greig Hazard Research Centre, 1998.

[10] Vita-Finzi, C, *Seasonal Enhancement and Seismic Triggering by Impacts*, Royal Astronomical Society/Geological Society/British Interplanetary Society lecture defining the effects of subcritical impacts, 1998, pp. 3–9.

technology, feet and toes constantly damaged by frostbite and hardened by the drudge across permafrost and rough terrain; a nomadic
life in constant quest of food and sanctuary. Our ancestors' land
must have been beautiful, yet harsh, much as the Alaskan and Asian
plains are today, and so much of Eurasia and the Americas were
then. Huge lakes carved out of the land – in truth a footprint left by
the retreating glaciers on their journey to the north and south
extremes – teemed with fish and provided an abundant new food
supply for the adaptable Cro-Magnon man.

As the glaciers retreated, so too did the great herds of glacial
animals, creatures that had been constant companions and sources
of food and materials for human beings during the 100,000 years of
ice. Once-common predatorial monsters, which might not have
looked out of place in the workshop of some fantastical Victorian
faker, became scarce. The giant sabre-toothed tiger, with its 15-
centimetre fangs that protruded from the jaw like great tusks, a
creature from our worst nightmares, for example – and yet our own
kind once fought a constant battle against it. Man hunting beast for
food, the beast retaliating and fighting for survival: how could we
face such creatures and survive? Running through icy grasslands and
frozen glaciated valleys with little shelter and only their wits to
protect them, our forefathers not only faced these monsters but
hunted and killed them, too. They proved, generation after generation, that the hand and mind of the skilful hominid was more than
a match for the monstrous beast's raw power.

The sabre-toothed cat was not alone in losing its natural habitat:
its prey also faced habitat loss. The cave bear and giant ground sloth
both competed, as the big cats did, for food in a changing environment, alongside the burgeoning population of humankind. The
prodigious hunting talents of the human may have been the main
killing mechanism for these monsters. Certainly, as the Ice Age ended
and the current interglacial began, approximately 10,000 years ago,
our ancestors developed new fishing techniques, using nets in the
glacial lakes, and developed advanced weaponry that was simply
devastating on open or forested terrain. This was a type of weapon
that could kill over tens of metres: the bow and arrow. Still in
practical use in some Third World nations today, this technology
must have developed as a direct consequence of changing environmental conditions which resulted in the emergence of forest-dwelling

56

game animals; sleek of limb and fleet of foot, these fast-moving targets were otherwise difficult to hunt and kill.

The great mammalian ice giants, especially the mastodon, mammoth and bison, must have found it increasingly difficult to thrive in an environment now populated with evergreen forests, rather than tundra-like grasslands, and permafrost. They moved slowly and had been easy prey for the human hunter. We were clever, probably working in teams sending volleys of stone-tipped spears into the great lumbering beasts, wearing them down until they were easily finished off. Just as whale hunters might use harpoons today, so we did on these vast land creatures. Small wonder that the mammoth and its other giant cohorts succumbed to the new inter-glacial hominid.

And so the absence of impacts and the general warming of the planet would have brought society and kinship, ever more inventive technology and, crucially, the development of farming and the concept of owning land. Worse still for the ice giants, the human groups were blossoming, their numbers growing, adapting more speedily to this new terrain, their keen problem-solving minds ever watchful for a new way to kill.

We developed in order to survive our world in constant upheaval, and only the keenest of bodies and minds could develop the skills necessary to thrive in a world of ice followed by a rapid great thaw. The thaw was a perturbation of constant climate, every bit as distressing to living organisms as the onset of the Ice Age itself. Our evolution, steered by constant climatic perturbation – thanks to volcanic activity, plate tectonics, Milankovitch cycles and, most importantly, bombardment from space – forced us to become the ultimate survivor of the last great epoch of ice. Yet problems lay ahead for the children of our kind who welcomed the warmth of that first great spring in a hundred millennia. No frozen world can thaw without consequence: indeed one of the greatest terrors known to man would lay its hand of death upon us – the flood.

Not surprisingly, impacts may not only be the delivery mechanism of the final hammer blow that throws Earth into ice ages: they may also end them. A mechanism similar to the dust-loading phenomenon that triggers cooling could also kick-start the greenhouse effect. It is not immediately apparent why aerosol debris should result in warming.

During the last period of glaciation, the average global temperature fluctuated between approximately −3 to +7 degrees Celsius. The atmosphere was correspondingly dry, with low evaporation and therefore little precipitation. The atmosphere was approximately ten times as dusty as our own. Despite these prevailing conditions, an all-consuming change took place: in probably less than 100 years, with average temperatures rising to 17–22 degrees Celsius,[11] high evaporation levels coupled with precipitation rates of 80 centimetres per year marked the beginning of the Holocene thaw. Why?

An impact event could change the Earth's albedo during an ice age sufficiently to induce the onset of the thaw. A scattering of ice particles derived from a comet of 2 kilometres in size, which it has been estimated could occur at least every 0.2–0.3 million years, into the upper atmosphere might release enough water to trigger global precipitation rates of 0.5–1 centimetre. This increase in precipitation and higher moisture content in the atmosphere as a whole would reduce the Earth's reflectivity (albedo) and consequently allow more solar radiation to penetrate and warm the surface.[11] The result would be a return to a more temperate Holocene-type interglacial climate, of the kind we enjoy today.

Impactors of this size have hit Earth many times in the past and millions may well have penetrated our oceans since the planet was young. These impactors may have triggered a renewal of the greenhouse effect, resulting in soaring temperatures, sea-level rises and, inevitably, floods. Others may have caused floods of biblical proportions simply by hitting the ocean and generating super tsunamis. Either way, impacts could literally be the progenitors of the great flood stories so common in human mythology.

It was discovered (by two British geologists, Scott Debbenham and Brian Storey) in 1997 that the Antarctic ice sheet, long thought to be stable, has in fact melted on several occasions over the last 5 million years, swamping Earth in floods of biblical proportions. The ice shelf on the western side of the 5,000-metre-high trans-Antarctic mountains is based on a rift – a series of fragmented islands – producing an inherently unstable ice sheet that is 3,000 metres thick

---

[11] Wickramasinghe, C, *Climatic Switches Induced by Stratospheric Dust Loading*, Royal Astronomical Society/Geological Society/British Interplanetary Society lecture defining the effects of subcritical impacts, 1998.

and stretches for 4,000 kilometres. There is enough water stored in that great frozen reservoir to change the surface of the globe. It is quite possible that this shelf melted 3 million years ago, producing a vast flood of the kind described by Ovid, the evidence for which has been found in sedimentary layers in New Zealand.

Gregory Zielinski, of the University of New Hampshire (USA) and part of the Greenland Ice Sheet Project 2 team, has speculated that an apparent increase in volcanism (demonstrated by the presence of sulphur and volcanic glass in ice cores) in the first 1,000 years following the great thaw from the last ice age (circa 10,000 years ago) was due to the loss of pressure exerted by the ice on the Earth's crust. He likens this to a 'popping cork'.[12] This popping effect may account for the volcanism that is believed to have caused Antarctica's western ice shelf to melt. Could an impact trigger a similar popping-cork effect? It is possible that impacts have actually accelerated the rate of plate-tectonic movement as well as altering the direction[13] of these continent-bearing structures. Thirty-nine inexplicable tectonic events occurred in the Phanerozoic period of geological history[13] (600 million years ago), which seem to tally with terrestrial and oceanic impacts, strongly suggesting that impacts are influencing the very shape of the Earth's surface.

If an effect analogous to the popping cork could be generated by impacts – collisions powerful enough to alter the course of tectonic plates – then the suspected volcanic plume thought to be responsible for melting the western Antarctic ice shelf might be generated again. It could be possible therefore that a major southern-hemisphere impact – a particular astronomical blind spot where no NEO (near-Earth object) observation programmes are currently operating – could drown Earth in an unimaginable sea of water.

Is there a precedent in geological history for an oceanic impact, and could these events have been caused by collision with nonterrestrial objects? Well, new evidence (craters and sedimentary clues) suggests that they could. Impacts could have generated volcanic plumes which released superheated rock (magma), raising sea temperatures by as

[12] 'Reading the Fine Print of Climate History in Greenland's Ice', National Science Foundation, PR: 94–56, September 1994.
[13] Price, N, *Evidence for Impact as a Significant and Periodic Geological Process*, Royal Astronomical Society/Geological Society/British Interplanetary Society lecture defining the effects of subcritical impacts, 1998.

much as 5 degrees Celsius globally – enough to melt the ice sheets, thereby raising sea levels by 2 metres worldwide, flooding many areas, including the United Kingdom.

Only one proven oceanic impact site has been studied with great interest since 1960. Located in the eastern Pacific sector of the Southern Ocean is an area referred to as Bellingshausen Sea – after the great Russian explorer of the Antarctic, Fabien Gottlieb von Bellingshausen (1778–1852) – west of Antarctica and east of New Zealand. This area saw a catastrophic impact 2.16 million years ago by the asteroid Eltanin. A recent mission, aboard the FS *Polarstern* in 1995, set out to reconstruct the events surrounding this vast impact, which released an estimated 100,000–1,000,000 megatons of TNT equivalent. The resultant paper, published in the scientific journal *Nature*,[14] suggests that the impactor was up to 2 kilometres in diameter – of a size sufficient to have globally destructive effects. Earth was already in the grip of the Pliocene glaciation phase and the impact generated a 'mega-tsunami' that may well have taken ocean sediments, rich in marine fossils, and deposited them on shorelines around the world.

In Pisco, near the Peruvian shore, geologists have found a chaotic jumble of land and marine skeletons in the sediments of the coastline. The remains appear to date from the Pliocene period and would tally with a vast deposition of marine material caused by the Eltanin asteroid's resultant tsunami waves, which were probably 20–40 metres in height and would certainly have reached the shores of South America and beyond. Interestingly, Antarctica has a number of reported marine fossils in its interior which conventional wisdom struggles to explain. The team aboard the FS *Polarstern* speculated that the vast explosion following the impact could have scattered marine debris around the globe, including delivering materials to the Antarctic interior, thus explaining, at least in part, the erroneous fossil presence. It is evident that this colossal impact, of a type that is thought to strike once every 2–5 million years, triggered vast oceanic disturbances, flooding areas as far apart as New Zealand and Peru. It is likely that other similar events have happened in the past – though finding evidence without an apparent

[14] Gersonde, R, et al, 'Geological Record and Reconstruction of the Late Pliocene Impact of the Eltanin Asteroid in the Southern Ocean', *Nature*, Vol. 390, No. 6658, 27 November 1997, pp. 357–63.

crater is extremely difficult – and will continue to do so at regular intervals. If so, the consequences could be catastrophic in a biblical sense.

Ovid's reconstruction of perhaps the definitive flood, which might be the origin for the great biblical flood overcome by Noah in his mighty Ark, has been preserved in mythological history by many great ancient scholars. Ovid wrote:

He [Neptune] sent forth a summons to the rivers . . . The rivers . . . went rushing to the seas in frenzied torrents . . . Neptune himself struck the Earth with his trident; it trembled, and by its movement threw open channels for the waters. Across the wide planes the rivers raced, overflowing their banks, sweeping away in one torrential flood crops and orchards, cattle and men, houses and temples, sacred images and all. Any building which did manage to survive this terrible disaster unshaken and remain standing, was in the end submerged when some wave yet higher than the rest covered its roof, and its gables lay drowned beneath the waters. Now sea and earth could no longer be distinguished: all was sea, and a sea that had no shores . . . The ocean, all restraints removed, overwhelmed the hills, and waves were washing the mountain peaks, a sight never seen before . . . So when the earth, all muddied by the recent flood, grew warm again, under the kindly radiance of the sun in heaven, she brought forth countless forms of life. In some cases she reproduced shapes which had been previously known; others were new and strange.[15]

Remember that Ovid wrote his *Metamorphoses* in AD 8, 2,000 years ago, basing them on Greek myths already thousands of years old. It is entirely possible that the Greek form of the flood myth is far from the original: it perhaps stretches back into an antiquity before the written word had been invented. Some would argue, however, that the flood myth, clearly retold in the Old Testament, dates from 2354–45 BC.[16] Dendrochronological evidence, from the Irish oak, displays distinct, even extreme, narrowing of the growth rings in the period dated to 2345 BC. Dr Michael Baillie of Queen's University, Belfast, perceives this as a milestone in history, 'a marker date', as he puts it. By that Bailey means that cross-discipline study

[15] Innes, MM (translator), *The Metamorphoses of Ovid*, Book I, Penguin Classics, 1955, pp. 39–43.
[16] Baillie, Dr Michael, Queen's University, Belfast, interview, 1998.

produces volcanic evidence from the Hekla 4 volcanic eruption of approximately the same period, but more keenly it is seen as the date of catastrophic flooding, and he points out that legend decrees that Chinese history began in 2345 BC.

Others, including the famous British archaeologist Charles Leonard Woolley (1880–1960) – an original archaeological adventurer who would not have seemed out of place in the pages of the *Boy's Own* adventure comics of the 1930s – date the great flood to 3500 BC. Woolley became famous for his excavations at Carchemish in Syria, Tell el Amarna in Egypt and the ancient Mesopotamian (now Iraqi) city of Ur. Woolley popularised archaeology for the first time, publishing accounts of his adventures and discoveries in an exciting format, spawning a whole new literary genre, which is currently enjoying a renaissance. Woolley's date for the great flood was based on his knowledge of the ancient texts and legends which originated in the cradle of early civilisation, a terrain that was literally 'his' intellectual and physical domain for many years.

Was it 2345 BC, 3500 BC or some other as yet undetermined date? I think it unlikely that any one date could be assigned to the great inundation myth referred to in the ancient texts and legends. There have been no fewer than 1,500 floods recorded by scribes in the last 3,500 years in the Huang He (Yellow River) region of China alone. I believe that the accounts, so familiar thanks to the Old Testament prophets and the much older pagan myths, address a series of constantly recurring events – vast in their import to the environment and extraordinary in their violence – which have threatened human history since the beginning of the Holocene and end of the Pleistocene Ice Age.

The floods conquered by Noah – or, in Ovid's story, survived only by the Greek equivalent of Noah, the virtuous Deucalion, and his wife Pyrrha – may well owe their genesis to fragments of a vast comet penetrating the ocean and causing tsunami tidal waves. It is also possible that a partial melting of the unstable western south polar ice sheets, caused by a short-lived global-warming effect, created a huge increase in water volume. This would be possible if the comet debris hit calcium-carbonate deposits on the sea bed, throwing large amounts of carbon dioxide into the atmosphere and creating a short-lived, intensified greenhouse effect – as seen today, thanks to modern pollution levels. This could have resulted in a great increase in precipitation and sea levels:

> The waters of the flood were upon the earth ... were all the fountains of the great deep broken up, and the windows of heaven were opened. And the rain was upon the earth for forty days and forty nights ... And the waters prevailed exceedingly upon the earth; and all the high hills, that were under the whole heaven, were covered ... And all flesh died ... the waters prevailed upon the earth an hundred and fifty days.[17]

Alternatively it is possible that a comet impact released enough water into the atmosphere – perhaps even as much as 5×10^{18} grams[18] – to generate elevated precipitation alongside a mega-tsunami and a disastrous shockwave.

This sort of impact shockwave might have been recorded in Ovid's version of the legend as being the actions of the sea god: 'Neptune himself struck the Earth with his trident; it trembled, and by its movement threw open channels for the waters.' Clearly a catastrophic earthquake shattered the land, ahead of the coming tsunami; indeed an oceanic impact could not be better described than in Ovid's *Metamorphoses*. More recent tsunamis such as that which swamped Lisbon during the 1755 earthquake may also have been triggered by a colossal impact with the ocean. It seems likely that these myths are accurate records of archaic knowledge. Indeed the reference in Genesis 6:4 that 'there were giants in the earth in those days' does make one wonder what the Old Testament prophet (said by some to have been Moses himself) meant. Could he have been referring to extinct species of mammoth and other great land animals, lost because of the great flood? It is unlikely that another hominid group would be referred to as 'giant': the 75–80 skeletons found to date reveal Neanderthal to have been very muscular but probably somewhat shorter than *Homo sapiens sapiens*. If Neanderthal survived much later than 35,000 BC, perhaps even to the beginning of the Holocene, it is possible that his few remaining communities were

---

[17] Genesis 7:10–24, King James Bible.
[18] Wickramasinghe, C, *Climatic Switches Induced by Stratospheric Dust Loading* Royal Astronomical Society/Geological Society/British Interplanetary Society lecture defining the effects of subcritical impacts, 1998.

swamped by the numerous floods that must have occurred following the rise in sea levels triggered by the great thaw.

The single most perplexing question is how we, the human race, survived these great floods at all. We have seen, with the Hurricane Mitch catastrophe of 1998 for instance, that as a species we are pretty useless when it comes to dealing with severe flooding: many thousands died deaths that were not, on the whole, due to the winds themselves, but to the flood waters and the mass of muddy slurry washed from the mountains of Central America. How then did we survive the great inundation? If the theatre of these biblical catastrophes was on a scale greater than that typical to more commonplace regional flooding events, and the world was truly swamped for extended periods (as Baillie's tree-ring data would suggest for the 2345 BC inundation episode and Woolley's hypothesis intimates for the 3500 BC occurrence), just a few pockets of Deucalion-type survivors would have been responsible for repopulating the human race. An intriguing thought.

Indeed it is a feather in the cap of all creationists that the Human Genome Project, a research venture, established by the US National Institute of Health and Department of Energy in 1990, to map the entire complement of human genetic characteristics, has produced enough evidence to prove that all modern humans descended from the same group of hominids. The 'Out-of-Africa' theory and the inundation myths seem to have been given a ringing endorsement by this discovery. I have to conclude that there may be some metaphorical connection to this 'genetic fact' in the genesis legends common in all societies around the world. Adam and Eve first begat man. Noah was a tenth-generation descendent of the original big-daddy progenitor Adam and it was from this stock that the rest of humanity was repopulated. In the Greek version of the legend Deucalion and Pyrrha were the bearers of the human gene pool. If we are all descendants of Noah, why is it that we live for only three score years and ten when Noah himself was 600 years old at the time of the great flood? Perhaps the inbreeding needed to repopulate the species (a topic barely discussed by Old Testament enthusiasts but offering moral theorists ample scope for debate) produced this genetic mutation, cursing us to fleeting life spans – when compared with the great Ark builder, at any rate.

It is likely, I believe, that impacts were responsible for these events

– the descriptions do not concur with volcanic activity, and their severity, in some cases preceded by shockwaves, demands an extra-terrestrial origin: a deathrock.

We live in a period fraught with millennial preoccupations, an occurrence common in history after 1,000 years of progress. Our modern concerns tend to focus on the preservation of the environment and world peace. Global warming is an issue seldom far from the headlines, and yet ironically this interglacial period forms part of an unusually cold period in Earth's history. Remember that the Milankovitch theory and the dust and impact cycles all seem to tally with a 100,000-year time span for global ice ages. The last ice age ended nearly 10,000 years ago. The next is imminent and could hurl us into another 100,000 years of ice. Alarm bells ought to be ringing, yet the silence is deafening. Is it not time for us to consider the consequences of such a massive and effectively eternal (in terms of human history) climatic shift? A world subjected to a biblical flood is likely to be one devoid of most animal life. A world lost to marching glaciers and tundra, on the other hand, might well be an environment that we could survive and, in time, thrive in. I believe that the house of cards laughably referred to as economy and society can withstand such a shock only if we begin to understand the challenges presented by eternal winter and develop a new socio-economic paradigm to deal with this potential threat.

What if we were thrown into an ice age by bombardment?

An impact winter gone mad, stretching through the centuries, changing our land and our lives, effectively for ever – ten times longer than recorded history. Great ice sheets would grow from the poles, spreading across the continents. At first the planet would be quite warm; fogs would be common but severe, making transport difficult. Dust in the atmosphere would escalate to ten times its current levels because of the impact and the increase in ice and snow particles in the atmosphere, just as it did in the last ice age. By the end of the decade after the onset of the ice, air transport would become extremely hazardous, though the fog would decrease as global temperatures dropped.

Agriculture would be thrown into turmoil – crops would suffer frostbite and wither as photosynthesis declined. Cattle would starve as once-lush grazing land froze, leaving little more than tundra behind. Farming habits would have to change, requiring a greater degree of husbandry to help grazing animals survive the transition.

But with grain in short supply, thanks to the great freeze, the enormous population of husbanded herds common today would have to be reduced in order to guarantee viability. As time progressed, the climate moving firmly into an ice age, the food chain would change, as species common today might find it almost impossible to survive. Even the common household pet might fall prey to food shortages as it became a luxury that few would be able to afford.

Fuel crises would be likely, initially because of the transport problem, worsened by the increasing difficulty in excavating and drilling for oil in the new environment. Demand for fuel stuffs would greatly increase, as overworked generating systems consumed vast quantities in order to provide heating and lighting in this frozen world. How long before the infrastructures failed to meet that demand?

Accidents – in the home and workplace and on the roads – would increase, thanks to poor visibility, treacherous conditions and failing health. Systems of health and policing would be correspondingly strained, might even collapse, owing to the additional demand for their services.

As time progressed, the average population age would become younger as the older members of society succumbed to terrible conditions and worsening diets – a problem intensified by the rising cost of fuel and food. The shortage of food, worsened by transport problems, would have to be addressed by the adoption of a planned economy and most especially the introduction of rationing.

Day-to-day living would be hellish. Blocked roads, road rage and petrol rationing would result in less travelling, and those that did travel would no doubt face journeys fraught with danger. Trains might be more effective than road vehicles, though it is unlikely they would be able to meet the whole demand from workers and commuters. Once the road network succumbed, commuting would become very difficult, travel expensive and less common.

Teleworking might offer an initial reprieve for some industries, though such a great increase in electronic traffic is likely to slow down email and Internet servers considerably. Even now the Internet slows when America logs on.

Scheduled blackouts, reminiscent of those imposed during the European strikes of the 1970s, would be one solution to the energy problem. Coal might be one fuel source that would enjoy a renaissance in Europe, owing to its accessibility and plentiful supplies on the Continent and in Britain.

Depression and suicide rates would be likely to increase as people succumbed to seasonal affective disorder (SAD), a condition in which people enter a depressive state when moving from one season to another – usually summer to winter. This would be a winter without end and might seem like a winter without a hope of reprieve.

Work, income, food, health and travel problems, depression and sickness, are likely to result in a significant increase in social problems, particularly alcohol and drugs abuse. Violence is likely to increase – most domestic violence (70 per cent) occurs when the male cohabitor is drunk or under the influence of restricted drugs.

What of the economy? A shock to the world markets, then a rallying followed by a slow decline – such has been the response to natural disasters such as Hurricane Mitch in 1998. So we might reasonably assume a similar response to predictions of, and the onset of, an ice age. Further than that it is difficult to even speculate. It is likely, however, that world markets would decline slowly as industrial production slowed. As production falls so income falls; income depression results in a fall in demand for goods and services and has a downward pressure on prices. This downward price pressure may be mitigated by the likely increase in the cost of raw materials (owing to transport and excavation problems) to the manufacturer.

The futures and insurance markets would collapse, as confidence in the long-term equilibrium of the world economy, balanced by depression followed by boom, would be mitigated by doubt about how events would unfold. No one in recorded history has lived through an ice age; how are we to be certain what to expect? How can we know the way in which society, industry and our economy will behave in these circumstances, in uncharted waters? Economics is built upon a basis, as with all sciences, of empirical evidence, past cycles and established norms. If the world enters an era so alien to the one it left behind, the science of economics would fail; no accurate predictions of future prospects could be made.

The world's stock markets may eventually suspend trading, as they did in Russia in 1998, until the world downturn bottomed out (i.e. reached the base of the downturn and industry began to deal with the new set of problems). Such an effort might be virtually impossible. Nevertheless, it is conceivable that in time human ingenuity would overcome the severe conditions. But how long would that take? Decades? A century?

Nationalisation and a planned recovery would seem likely, as has occurred twice this century in times of extreme duress (World Wars One and Two). Priorities would be fuel and transport, and once raw materials could be moved from the point of extraction, then production could be stepped up once again and geared to the new climate.

Morale would have to be managed by nation-state governments, perhaps using television and radio and other media, working in much the same way as propagandist films, songs and posters did during the World Wars. An active plan, with a well-defined agenda of achievable targets, would help to rally nation-states. The siege mentality would probably be encouraged: 'work for the good of all' might be the sort of phrase that would return to the popular agenda. Similarly, investment would be vital in order to secure industrial recovery; therefore something analogous to the war bond might help to stimulate investment in national economies. With travel limited by environmental conditions, the idea of community might also enjoy a renaissance – community-care centres might well be encouraged to help provide food and medical help and bolster morale.

If trade routes could be reopened and maintained, then the major stimulus might be the recovery of the American economy. An international effort to inject aid into severely troubled economies and nations would be difficult at first, though practical help would be a priority in order to stimulate international trade, much as the 27 'Paris nations' attempted to put together an economic rescue package – mitigating debt – for Central America in the wake of Hurricane Mitch. The Paris group of creditor nations suspended debt repayments for three years in the hope that the Central American economy will not collapse and damage world trade as the unstable Russian and tiger economies threatened to do in late 1997 and 1998. Although Hurricane Mitch initially killed approximately 11,000 people and cost Honduras alone £2 billion, the damage done pales when compared with the devastation likely during an impact-precipitated ice age.

Depending upon the severity of change, practical planned efforts might mitigate the effects of the onset upon our economy and society, but it seems unlikely that things would continue as we know them, untouched by the winds of change.

The horrors of Hurricane Mitch should act as a good and timely reminder that waters can flow very easily, and so too can ice; both could encroach on our civilisation, seeping into the weak spots and cracking it apart. A terrible shock, triggered by an impact, can occur without warning tomorrow, next week, or in 52,000 weeks. Yesterday's events will happen again tomorrow; the ice will return.

CHAPTER

FOUR

TOMORROW IS YESTERDAY

An Irish Republican Army bomb threatens the Northern Ireland peace talks... A milk churn explodes in Belleek, Ireland... Were years of diligent work, including many political sacrifices by the Loyalist and Republican movements, to be threatened by a rogue bomb? No. The milk churn exploded, but its destruction had a far less terrestrial origin, as this newspaper report revealed on 8 February 1998:

Meteorite, not IRA, caused blast

astronomer solves mystery of explosion in Fermanagh

The blast ripped through Belleek in Fermanagh at 5 a.m., terrifying locals. They suspected it was another IRA bomb at first, except for old Mary Rose Doogan. She thought her gas canisters had exploded... The police and army were on the scene within minutes... They carried out an intensive search and pulled in reconnaissance aircraft to photograph the area. They found nothing, and were left flummoxed... Four weeks later a farmer looked behind a hedge and discovered a crater, nearly two metres across and one metre deep... The security forces thought the shattered milk churn nearby was proof of a bomb... That it was found close to a stream running underneath a main road – a favourite planting spot for the terrorists – seemed to seal it... But the explosion which hit Belleek last December was, it seems, a one-tonne meteorite, the first [known] to hit Northern Ireland since 1969... It is thought to have come from the asteroid Phaeton[1]... Tom Mason, director of the Armagh Planetarium, is leading the investigation... He has bought the... churn... so that he can take custody of the punctured can when it is returned from the forensic laboratories in Belfast... Doctor Mason suspects the meteorite exploded into millions of pieces on impact... His breakthrough came on Wednesday when he discovered stuck to the churn a small particle of glassy rock... He

[1] 3200 Phaeton, a small body with an unusual orbit.

70

is convinced that the tiny fragment was formed on impact when the hot rock hit the churn and exploded.[2]

The peace talks survived this threat. It is clear that our evolution and the development of our society are not the only elements beholden to the impact phenomenon. Ancient history is replete with catastrophes, whether terrestrial, such as Krakatoa in 1883, or celestial, such as that described by the Phaethon myth, but the ancients were not the only humans to be influenced by the 'celestial sphere' as they knew it. Countless times this millennium the course of recent history has literally been changed by the presence of bodies from space. Are we prepared to allow the possibility of catastrophe to haunt our future as it has our past? Not all outcomes have been bad – our society as it is today is at least in part the product of celestial events. The bombardment from space 4,500 million years ago provided many of the essential building blocks of life – a generally accepted fact – whereas all history, evolution and even our climatological environment depended (and still does depend) upon impact – a rather more contentious assertion.

Do we perceive our current physiognamy and society to be the ultimate form, the pinnacle? If we allow impacts to kill, influence the history of the future, perhaps even trigger mutations – as some Russian scientists believe – our society will inevitably change because of it, if we survive the impact. If we decide to face the threat, understand and conquer it, we will then sever one of the age-old symbiotic relationships between Earth and space. We will literally take charge of our destinies in a way that the dinosaurs could never have achieved, despite their unrivalled longevity.

Some people ask me why we should be so worried about the impact threat – why not hurricanes, earthquakes, nuclear war? It is a matter of scale: the impact threat is more destructive than any other known phenomenon. Short of some vast interstellar catastrophe such as a supernova, nothing in the known galaxy 'out-guns' this force.

If we could take charge of this situation, move our slumbering giant of a world socioeconomic structure and stir it to produce an international rescue package for the future, we could conquer any problem: social, political, physical.

[2] Mullin, J, Ireland correspondent, 'Metoerite, Not IRA, Caused Blast', *Guardian*, 8 February 1998.

If.

I will address the practicalities of improved observation, detection and potential solutions to ameliorate the threat in later chapters. That said, there is a model that we can look to for inspiration, one that offers hope for international cooperation on a scale that would be necessary to avert such a cosmic threat: the new international space station – Alpha.

The station stands out as a milestone achievement, fifteen nations working side by side, both financially and physically, Russians and Americans included, for one common goal. The five-year (1998–2003), 45-flight orbital assembly schedule of the new space station is itself a marvel and a reflection of the level of sophistication that our space technologies have achieved. The single most important fact about Alpha, apart from having a permanently manned presence in space, is that we – *Homo sapiens sapiens* – can now justifiably claim to be capable of working collectively. In a way we are beginning to understand the value of the insects' collective efforts. Not just on some international agreement about arms or trade, but to physically integrate disparate technologies, language, concepts and cultural norms. The station may indeed be the most elaborate civil-engineering project ever undertaken by humankind, but it is also the first real step, in every field of human endeavour, in the direction of globalism.

Space research has always contributed to that effort, ever since Yuri Gagarin (1934–68) first looked down on Earth from space in 1961, buckled into his minuscule Vostock capsule, and saw a small but mesmerising blue-green jewel in the darkness. His eyes, human eyes, saw what none of us had seen until that moment – that our world is not some impervious gigantic universe in itself: it is a precious gem in a setting rich with billions of other gems, each different, each fragile and open to exogenous onslaught from space-borne debris. It was a sight so unsettling, so astonishing, that Gagarin never truly recovered from his experience. He was not alone: many astronauts find their whole attitude to life changes, as though looking back upon Earth as only they can were a spiritual event. Their insight into the fragility of our world is remarkable and precious.

Space station Alpha may serve to further this understanding of our place in nature. Our lives are not detached from space: we are moving *through* space – passengers on a great living starship which races through the cosmos on an orbit that carries it around a sun, which is itself part of a greater system that is travelling through a galaxy of such assemblages on an orbit that takes 250 million years to complete, in a universe of galaxies. We are part of that vast complex – it is our home, and so can be affected by the objects with which we share our galactic habitat.

Any anti-impact scheme would rival the international space station in scale, perhaps even dwarf it, but could use Alpha's development as a political, financial and physical model for the construction of an international impact-defence agreement and facility.

You might be forgiven for believing that fear and ignorance of the kind experienced by the pagan civilisation of Montezuma II's Aztec Empire has little analogue with modern Western society. That would be an understandable but totally erroneous conclusion.

Montezuma II (1466–1520) reigned from 1502 to 1520. He was the last ruler of the great Aztec civilisation. A tremulous and indecisive leader, he was obsessed by omens and magic, and was haunted by personal 'demons'. His fascination with mysticism, however, led to the downfall of perhaps South America's greatest civilisation. Why? Montezuma II saw a great portent of doom, a celestial warning of his downfall; he seemed to believe that everything he did was foretold and that it would come to nought. He saw a great comet which broke into three in the 'heavens' creating trails of fire across the sky.

In 1519 a flotilla of twelve ships carried the white-skinned would-be gods to Montezuma II's domain: Tenochtitlán. The Spanish conquistadors, intent on establishing a South American empire, were invited into the king's court. Petrified of their strange appearance, afeared of their 'manifestation' aboard vast, apparently ethereal ships, heralded by the exploding comet, Montezuma II offered no resistance, but vain attempts at trickery and black magic. The conquistadors, led by Hernando Cortés (1485–1547), put Montezuma under arrest and took his vast kingdom. Within eighty years, nearly a quarter of a million Spaniards had colonised South America. The old Aztec Empire lay in ruins, its buildings torn down, its customs abandoned. Why? Because of an apparently portentous celestial omen and one foolish man's fear and inactivity.

Montezuma II was not alone. Nearly 500 years earlier another pivotal moment in history was decided, in part, because one great leader feared defeat due to another 'celestial portent of doom': the date was 1066; the place, England.

Harold Godwinson, brother-in-law to Edward the Confessor, fought a long political battle to stake his claim as the next rightful heir to the throne of England. Two powerful rivals, Harald Hadrata of Norway and William, Duke of Normandy, both laid claims to the English crown upon Edward's death. A bloody battle and ultimate victory for Harold, waged at Stamford Bridge, Yorkshire, repulsed Hadrata's claim. Acting swiftly, William of Normandy launched an armada of ships towards England, arriving on the South Coast but two days after the Battle of Stamford Bridge.

Harold's forces were battle-weary, though joyous in victory; they were in truth in no condition to repulse another contender. By mid-October, Harold had mobilised his forces once again, making his stand against the French forces, under William, near the village of Hastings, Kent. On 14 October 1066, William of Normandy's men and Harold's battle-scarred army met in a terrible campaign. Nearly all of Harold's knights fell, as the King fell. After travelling to London, William the Conqueror, as he became known, was declared King of England on the symbolic day of 25 December.

Why, you might ask, did a battle-proven force, sovereign to the country, fall to a small aggressor army? There is good evidence that Harold – crowned King of England in January 1066 – was convinced that he would fall in battle that year because of the appearance of a frightful omen. The *Anglo-Saxon Chronicle*, a record of English history prior to William's conquest, kept by monks and thought to have been initiated with the consent of King Alfred (849–99), records the Eastertide event in some detail:

> Then was over all England a token seen as no man ever saw afore. Some folk said that it was the comet-star . . . It appeared first on . . . the eighth day afore the calends of May and shone all week.[3]

The portentous presence was also recorded by the seamstresses of the Bayeux tapestry, a vast pictorial scene recording the history of William's ascension to the throne, created under the guidance of William's stepbrother, the Bishop of Bayeux.

The comet was no harbinger of doom for William, yet men of superstition and irrationality might fear an object if they have developed a belief that it will cause harm. It seems not unlikely that

[3] Modern translation courtesy of the British Library, 1998.

the oral tradition that comets cause harm, probably dating from the long-forgotten tales of witnesses to an impact, so affected Harold.

These pivotal moments in history were steered by the appearance of comets, and this particular event is known to have been the return of a comet later identified by England's Astronomer Royal, Sir Edmond Halley (1656–1742), the one that visits Earth with a periodicity of 76 years: Halley's Comet.

These great leaders allowed the presence of comets to change the path of history. Another leader, much less distinguished than these two great figures, led his followers to their graves thanks to an ominous cometary portent. This leader did not fall at the beginning, nor the middle, of this millennium, but at its end: March 1997. The place: Rancho Santa Fe, California, USA.

A bright, new rogue or long-period comet (with an orbit that carries it away from Earth for more than 200 years) was detected independently by two American amateur astronomers, Thomas Bopp and Alan Hale, in 1995, and measurements of its albedo produced estimates of its diameter amounting to 40 kilometres. This vast comet made its closest Earth approach on 22 March, 1997: 1.315 Astronomical Units away.[4]

As the comet passed Earth a group of 38 fanatical cultists led by Marshall Applewhite, a 66-year-old retired music teacher, sent electronic messages around the world telling their followers that the time to ascend to a higher plain had arrived: Comet Hale-Bopp was a herald of their coming rescue from Earth. The group were known as the Heaven's Gate cult.

Newspaper reports and wires carried bizarre video images and transcripts of the group's farewell decree. They spoke of their belief that the Earth's atmosphere was to be recycled and only by ridding oneself of the 'flesh' could one ascend to a spacecraft which, they believed, followed the great gas and ion tails of Hale-Bopp.

Survival requires that you allow nothing of this human existence to tie you here [wrote one of the group, who was known as Anlody]. No wealth, no position, no prestige, no family, no physical pleasure, and

[4] 'Hale-Bopp: the Great Comet of 1997', PR, NASA Jet Propulsion Laboratory, 1997.

no religion spouting to hang on to any of the above will enable you to survive. They are only entrapments.[5]

This group was obsessed with fear, of the kind experienced by the pagan civilisation of Montezuma II and even by an eleventh-century Christian King of England; fear of a dreadful event, either brought by a comet or warned of by this space-travelling ball of ice and stone. It seems incredible that a group of modern individuals resident in California, one of the world's most economically and technologically advanced areas – home to Silicon Valley – could harbour a group so afraid of some demonic destruction that they took their own lives in a mass suicide in order to escape its coming. Yet it happened. Surely, education and discussion, politicisation of this issue, must be seen as important in light of such an event.

They were not alone. Others this century have seen fireballs that really were omens of bad times – omens that brought terrible destruction – the first of which occurred while my grandmother celebrated her sixteenth year of life: 30 June 1908. This blast, caused by a bolide exploding in the atmosphere, was nothing less than a sign of Armageddon for the Siberian Inuit tribesfolk of the Tunguska region. One old tribeswoman spoke of her memory of the incident: 'When the meteor fell, the whole Earth shook. The sky turned red, like flame or fire. It was like a rainbow passing overhead.'[6]

The Inuit feared for their lives and with good reason: an area of 2,000–2,500 square kilometres was destroyed by the blast. Their reaction was not mirrored in the rationalistic West – nothing frightened the educated classes. *The Times* of London was, in 1908, a well-known forum for discussion of events, its news content largely overshadowed by the letters sections, which were given much greater prominence than in modern versions of the newspaper. One *Times* reader wrote, on 3 July:

> Struck with the unusual brightness of the heavens, the band of golfers staying here strolled towards the links at 11 o'clock last evening in order that they might obtain an uninterrupted view of the phenomenon. Looking northwards across the sea they found that the sky had the

[5] 'Cult Leader Among Dead at Home; Autopsies Begun', Associated Press/ Nando, 1997 (quote unabridged).
[6] *The Day the Earth Was Hit*, Channel Four, 1997.

appearance of a dying sunset of exquisite beauty. This not only lasted but actually grew both in extent and intensity until 2.30 this morning, when driving clouds from the east obliterated the gorgeous colouring. I myself was aroused from sleep at 1.15 a.m., and so strong was the light at this hour, that I could read a book by it in my chamber quite comfortably.[7]

These two twentieth-century cultures not only lived thousands of miles apart geographically but possessed cultural views that were absolute opposites. One, the Inuit, still feared the unknown. Theirs was a fear born of first-hand experience, their lives dominated by the realities of survival, not yet tempered by the luxuries of Western-style, First World market economics. If the Inuit of 1908 wanted food, they grew, bred, hunted or traded for it. Indeed, even today a visitor to the Tunguska region would be surprised by the Inuit way of life: the small groups still gather mushrooms in summer and dry them on lines in order to bolster their meagre diets in winter.

Theirs was an existence barely different from that of a late Bronze Age agrarian society. Tribal and living within a traditional micro-economy, the Inuit still understand that danger lurks in every forest, for theirs is still an untamed habitat, as it was in 1908. Their lack of scientific understanding led them to believe that the end of the world had come and, if the meteoroid had exploded a little closer to their settlement, for them the world would have ended. Instead thousands of reindeer and dense forest were destroyed.

The educated classes of Western Europe had a totally different reaction: fascination tempered by complacency. They had nothing to fear, as Western culture dictated to them that ours was (and still is) a society where there is little to harm us. We live, as they did, in a tamed environment of paved concourses, street lights that banish the darkness, food in an urban environment that is bought, not, on the whole, grown – even in 1908.

Western science, driven by the uniformitarian doctrine, has achieved remarkable and vital breakthroughs. Rationalising celestial visitations is essential, but, while rationalisation is a cornerstone of science, as a society we have taken the dangerous step of perceiving a rationalised threat as a nonthreat: we have lost our sense of fear.

Not all Western groups felt as the readers of *The Times* did about

[7] Ingleby, H, letter: 'Curious Sun Effects at Night', *The Times*, 3 July 1908.

the eerie results of the Tunguska impact. Some social groups responded in a way not too dissimilar to that of the Inuit. My grandmother often talked to my father and the family of a time when her whole village, Hartburn, Stockton-upon-Tees in County Durham, was aggrieved by the terrible sights of the Tunguska night and, worse still, the 1911 visit of Halley's Comet.

> People were really afraid of [Halley's Comet]. I was. We thought the air would catch fire, that the forests might burn and the lakes and ponds would all dry up ... The townsfolk worried that they would go mad; I remember talking about it. We were worried that foreign armies would take advantage of the panic and invade – there was a lot of talk about invasions in those days, just before the Great War, especially from what we used to call the yellow peril: the Oriental people ... Lots of daft talk of snow in summer, raining fish and how we wouldn't be able to tell night from day. But we didn't know any different in those days. We were just frightened.[8]

The standard of education in industrial towns for working classes was poor – enough to allow the workers to fulfil their roles in the shipyards, mines and other heavy industry of the region, but little else. The Church and more importantly the local vicar still played a major role in their lives, as a social guide and general fount of all wisdom. Small wonder that they feared the visit of comets: their minds were filled with religious symbolism and warnings of eternal damnation in the fires of hell. Nevertheless, many of their ideas were not so very far from the truth: if such an object did impact, as in Tunguska, a fireball might well burn forests and the air, even turn night into day, as evidenced in London. Despite their raw fears, they perceived something of the truth. Yet their 'betters' had no such appreciation of the potential danger: they were obliviously content to leave such thoughts to scientists. Of course the Edwardian scientist was invariably too busy dwelling on gradualistic uniformitarianism to have any truck with the concept of catastrophic occurrences, however real they may have been. They had no paradigm for such things; therefore they did not exist – an approach that a modern psychoanalyst might declare to be possessed by those in a state of denial.

[8] Isabella Broadbelt, eyewitness account, 1989.

When did it become a good idea to abandon our fear? It is an emotion that has a valid function. I am not suggesting that we should live constantly under the shadow of fear, but that a healthy respect for the powers of nature cannot be a bad thing. After all, whose reaction to the Tunguska blast was more sensible: that of the mad-dog Englishmen playing cricket by the midnight luminance or that of the Inuit scared for his life – but a few kilometres from a blast that killed all animals and defoliated and flattened trees over a 2,000-square-kilometre area? If I were in a forest filled with dangerous predators, I would opt to follow the Inuit rather than a well-heeled City gent, my perception being that the man with a healthy fear might have a chance of surviving. The man who walks through a forest of carnivores in sublime ignorance is unlikely to complete his journey. We should fear these stealthy would-be impactors – they are potentially extremely dangerous, deadly even.

The people of the Heaven's Gate cult terminated their own lives in order to exit their 'virtual-reality bodies', as they put it on their video suicide declarations. The body was, to them, merely a vessel, not a true existence but merely a way of holding their 'life energy'. These people ran a successful business, a large hi-tech company that created custom-built websites for businesses. They used their own websites to recruit members and distribute their codes and beliefs – a ridiculous mixture of Christian scripture and UFO-abduction gibberish. They saw themselves as higher evolutionary beings waiting for their 'harbinger': Hale-Bopp.

Applewhite may have been slightly deranged, or simply believed what he preached, but his concepts are alarming to most of us. Astonishing figures can latch on to an event, such as the Comet Hale-Bopp, and become convinced that it is a sign of something pivotal and so make that apparent future event a reality: it has become a strange, self-fulfilling prophecy. Unfortunately, this sort of figure, a leader, has so often cost many millions of people their lives, or their way of life, simply by following such a strange and distorted view of reality.

Applewhite was referred to as Do and often appeared as the scholarly old gentleman, summoning his followers to embrace his ideology. In another video prepared before the mass suicide, Do tempted his followers by employing Christlike methodology and language:

I can be your shepherd . . . You can follow us but you cannot stay here

79

and follow us. You would have to follow quickly by also leaving this world before the conclusion of our leaving this atmosphere in preparation for its recycling.[9]

Men like Applewhite are perhaps as dangerous as autocrats and dictators in the Adolf Hitler mould. It is possible that an illness in the 1970s, reputed to be a heart attack, resulted in some brain damage. Certainly a rumour persists in journalistic fields that Applewhite had a near-death experience while in hospital and thereafter believed that he would ascend from Earth in a Christlike fashion, taking his followers to an alien spacecraft which followed in the wake of a comet. A dangerous and, sadly, deadly delusion. The cult committed suicide by imbibing phenobarbital and vodka – a deadly combination. Autopsies subsequently revealed that Applewhite and some of his key male followers were emasculated, surgically castrated, in an effort to further distance themselves from the 'ways of the flesh'. The Christian Church was greatly upset by these events. Applewhite's wholesale theft of Christian symbolism – ideas of ascension to a higher plane, spiritual rebirth and being the 'chosen ones' – left the Church greatly afraid that his cult would be confused with its own religion.

This terrible loss, 38 lives then and countless others since – suicides committed elsewhere following Applewhite's instructions via the Internet – is a shocking demonstration of the effect of celestial visitors to those of unsound or ignorant minds.

I say again: even comets and asteroids not directly set to impact with Earth can have an impact, socially, economically and, most importantly, emotionally and psychologically. Can we really afford to allow yesterday's mistakes to be repeated tomorrow?

Fear and ignorance, untempered by reason, nearly ended the lives of an entire Brazilian jungle tribe one morning in May 1931. The forest was alive with the sound of exotic birds and the fall of fishing nets into the river. Children no doubt played on the riverbank, helping their parents to fish for their breakfasts. A normal day, perhaps even an idyllic day.

Then the sky fell dark. A great burning object roared through the morning haze, a great ball of fire, its brilliant core encircled by a

[9] 'Cult Leader Among Dead at Home; Autopsies Begun', Associated Press/Nando, 1997.

fringe of spitting, boiling ears of flame, pushed back skyward by aerodynamic force, stretching out like a demonic tail. Then the asteroid succumbed to our atmosphere, the drag and incredible pressures too great for it to retain its form; it exploded into three great fragments. Together they raced over the canopies of the rainforest. The tribe must have watched in fear, having no comprehension of the coming terror. Finally, after travelling billions of kilometres since the beginning of the solar system, the great asteroid's three offspring exploded over the jungle floor. The sound was deafening, the light near blinding, as the equivalent of 50–100 kilotons of TNT was released by the three burning space rocks. The air burnt, the forest set ablaze by a blast equivalent to 4–8 Hiroshima-sized atomic bombs; 1,300 square kilometres of the tribe's home, the rainforest, was destroyed in a great creeping ground fire, a wall of flame consuming habitat and animals alike.

The villagers looked on in terror. Their acute senses, tuned to the forest, detected the death around them. A great fire of apparently celestial origin was consuming all that they knew. With their whole way of life threatened, and perhaps even their souls at risk, the local witch doctor gathered his people together and in an effort to appease the angry sky gods prepared to administer the deadly poison known as tempo.

A Catholic priest, a man of Western learning and extremely devout, had taken his doctrines to South America, fulfilling his apostolic obligation to the Church. Father Fideilia talked to the villagers and reassured them that their gods were not angry, offering them the olive branch of divine forgiveness for confessors of sin, under the Catholic doctrine. If this devout priest had not stumbled upon the tribe when he did, they might all have died by their own hand, in much the same way as Applewhite's followers did. Having visited the smouldering wastes where formerly there had been jungle, the priest recorded the events in his personal diary.

The fortunate happenstance of this priest's presence means that one more tale of destruction, of the kind witnessed in Tunguska, reached the First World this century. One cannot help but wonder how many other poor innocents took their own lives or were killed without warning, in remote areas, thanks to the impact of some unknown deathrock.

It is a sobering thought that both the Tunguska and Brazilian blasts were not strictly impacts at all, but were in fact atmospheric explosions, a few thousand feet above the ground; neither developed a crater, but both proved extremely effective destroyers. This lack of cratering is a pivotal fact when one is attempting to assess the level of risk from impacts. How many countless deathrocks have there been that we will never know of? Their regularity – once every 20–100 years – and their destructive force make them a truly dangerous threat. They are certainly of subcritical size and will therefore not generate a mass extinction, but they could easily destroy a city. The Tunguska blast pattern, the area devastated by the explosion, would comfortably smother London and its orbital road, the M25, with its butterfly-shaped form. It has been seventy years since the last known Brazil/Tunguska-type blast. The rapid expansion of urban areas and human population this century has resulted in countries with most or all of their people living in urban areas, such as Singapore (100 per cent) and Belgium (93 percent).[10] How long before a massive, densely inhabited area is devastated by an airburst from an exploding subcritical asteroid? The odds are that there will be such a disaster in the next quarter-century. No one knows for sure when it will come, or where it will hit, but it will happen.

What should have been the most frightening of all modern impact events – a portent of future human destruction – occurred in July 1994: the Jupiter collision. Discovered by the world's most active comet-hunting team – Carol and Eugene Shoemaker in association with David Levy – a vast comet that came to be called Shoemaker-Levy 9 had been torn into a string of giant fragments by Jupiter's gravitational field on a previous orbital pass near to the gas giant. The comet glowed like a row of burning beacons for sixteen months until their eventual destruction. These were real signs of celestial doom, great danger placards with 'Watch and learn, humanity!' written in large metaphorical letters along their frozen surfaces.

The great string's progress was closely watched, many scientists still doubting that such objects could do any damage whatsoever. US and world governments had thus far ignored warnings from influential scientists, such as the Shoemakers, and the computer simulations of Sandia National Laboratories. There was, as Eugene Shoemaker put it, 'a credibility problem'.

On 16 July 1994, the first fragment impacted with Jupiter's vast

[10] Eves, R, et al, *The Economist: World in Figures 1999*, Profile Books Ltd/ *The Economist*, p. 17.

surface. It tore into the Jovian atmosphere at speeds of 140,000–220,000 kilometres per hour, accelerated to such incredible speeds by Jupiter's vast gravity field. Jupiter's size is so great that its diameter is eleven times that of Earth's, its mass is 318[11] times greater than Earth's and it composes 66 per cent of the mass of the planets in the solar system – only the sun itself is larger. Despite the gas giant's great mass, Jupiter's hydrogen- and helium-rich atmosphere was devastated by the great shock of impact. The first fragment triggered an explosion so massive that its ejecta and gas plume rose over 3,200 kilometres above Jupiter's cloudbase.

Yet that first impact paled when fragment 'G' impacted. It generated an explosion that was so massive that its fireball was visible from Earth. Fragment 'G' left a debris cloud and melt sheet that caused atmospheric perturbation over an area larger than the surface of Earth. This incredible disruption to the giant of our system was no short-lived phenomenon, but remained cohesive and recognisable for almost an Earth year. The environmental consequences for Jupiter were massive.

Jupiter's atmosphere was severely damaged.

The post-impact reaction here on Earth was almost as spectacular: it fired the imaginations of my colleagues in the world press and the public alike. The media began to take the subject seriously and US politicians, ever watchful of the media, also began to understand the importance of research into the field of impacts. As we shall see, a number of groups began to develop methods of studying the phenomenon and, in 1998, NASA established a programme to detect near-Earth objects. The response to Shoemaker-Levy 9's impact was heartening, but only a beginning. The US government had responded to what they began to perceive as a threat. They were largely alone in this decision, however. Britain took no action whatsoever, its own political groups more concerned with internal division and fostering appearances of unity than taking action on anything that might be considered truly important.

The European Union was not so apathetic as to miss this awesome celestial omen. It acted, studied the phenomenon and suggested a way forward. The EU's declaration reads:

[11] Gribbin, J, *Companion to the Cosmos*, Weidenfeld & Nicholson, 1996, p. 233.

Action by the Council of Europe on the detection of asteroids and comets potentially dangerous to Humankind

There are two broad categories of space objects that have the potential to impact our planet: comets and asteroids. They are generally known among planetary scientists as Near-Earth Objects (NEOs). Their total population is unknown, but the number of Earth-Crossing Asteroids with sizes larger than about 1 km is estimated to be about 2,000. These objects are the most dangerous and only a tiny fraction of them have been detected to date.

Considering that the explosion close to the Earth's surface of even an object with a diameter of 50 m can have the effect of a 10-megaton nuclear weapon, the consequences of larger impacts would be disastrous on a global scale. The best-known, recent examples are the Tunguska ... and the violent impacts into Jupiter of the fragments of comet Shoemaker-Levy 9 (in July 1994); those fragments were only about 0.5 km in size, but caused devastation over a larger area than that of the Earth. Traces of other smaller impacts on our planet are frequently being discovered, as well as fossil records of cataclysmic impact events in the past.

The significant amount of information gathered over the last few years on asteroid and comet collisions indicates how they can trigger large-scale and long-standing ecological catastrophes, sometimes leading to mass extinctions of species; thus such impacts represent a significant threat to human civilisations ... the possible consequences are so vast that every reasonable effort should be encouraged in order to minimise them.

The Assembly therefore welcomes various initiatives ... [that take us] towards the development of a world-wide surveillance programme aimed at discovering all potentially hazardous NEOs and tracking their orbits forward by computer so that any impact could be foreseen some years in advance, allowing preventive actions to be taken as necessary.

The Assembly invites governments of member states and the European Space Agency (ESA) to urge the setting-up and development of [a] 'Spaceguard Foundation' and to give the necessary support to an international programme which would:

- Establish an inventory of NEOs as complete as possible with an emphasis on objects larger than 0.5 km in size;
- further our understanding of the physical nature of NEOs, as well as the assessment of the phenomena associated with a possible impact, at various levels of impactor kinetic energy and composition;
- regularly monitor detected objects over a period of time long enough

to enable a sufficiently accurate computation of their orbits, so that
any collision could be predicted well in advance;
- assure the co-ordination of national initiatives, data collection and
dissemination, and the equitable distribution of observatories
between northern and southern hemispheres;
- participate in designing small, low-cost satellites for observing NEOs
which cannot be detected from the ground, and for investigations
which can most effectively be conducted from space;
- contribute to a long-term global strategy for remedies against possible impacts.

Strasbourg, 20 March 1996

This report was, I think you will agree, a significant breakthrough
in political circles. I am a confessed federalist and would therefore
like to believe that this insightful paper, published only nineteen
months after the Jupiter collision (incredibly speedy by political
standards), will have significant impact on the policy decisions of the
EU member states.

Sadly, however, the wisdom stored in the Assembly seldom filters
down to the sovereign governments. Indeed, foreign ministers and
exchequers spend much of their time endeavouring to unpick any
resolution and rewrite it in favour of their own nation-state's
economy. Sadly, something as important as a space-originated catastrophe seems very remote to politicians who have, in all fairness,
many concerns over the management of health, education and other
domestic and international affairs. Nevertheless, politicians must be
made to listen to the declaration of the EU Assembly and, more
importantly, be encouraged to act upon it.

We are not dealing with the abstract or the obscure: we are considering the value of human life, the death of a city or a continent.
If those gatekeepers to power, the sovereign states' politicians, fail to
listen and understand the complexities of this issue, just as they did
with the much less significant BSE debacle, the cost will be measured
in ruined economies and perhaps tens of thousands or even millions
of deaths. The BSE crisis devastated Britain's beef market and had
severe consequences for the entire European beef market. Consumer
confidence fell to an all-time low. Yet the disaster also cost human
lives and may yet, some biologists believe, trigger an epidemic during the first two decades of the millennium. Despite these problems,
the BSE scandal, which has consumed copious hours of political

85

discussion in the British and European parliaments, is but a drop in the ocean when compared with the disaster that would be caused if even a Brazilian-type fireball (50–100 kilotons) were to explode over a city such as Manchester. Political inactivity would be an expensive mistake, economically and morally, by anyone's reckoning.

Remembering the understandable mistake British and Irish security forces made when confronted with the Belleek meteorite explosion, other recent revelations about tiny-impact events paint a grim picture for the safety of all our tomorrows. One of the most unsettling stories that I have come across revolves around a very poor home in a district of Bogotá, the capital of Colombia. It was a house full of children, no doubt throbbing with that peculiarly addictive atmosphere of excitement that only children seem capable of generating during the approach of Christmas.

On Sunday 14 December 1997, four children died alone and in agony. Living in a poor shanty house, devoid of power and light, the children were huddled together for warmth on a cold winter's night. While these innocents slept, a fragment of asteroidal rock hurtled through our atmosphere, friction from our air tearing at its surface, making it glow like a tiny, biblical falling star. The rock, now red-hot, hit their ramshackle home, turning its dry wooden structure into an inferno.

The local fire chief, Captain Carlos Augusto Rojas, reported seeing three fireballs. The ten-inch hole found in the corrugated roof left no doubt as to the identity of the culprit: it was clearly a meteorite. The fire department ruled out all other possible causes – there was no electricity supply and the house did not have candles or any other form of lighting. Other locals reported seeing fireballs in the sky prior to the house blaze. It seems that the four children had become the first in human history to be officially recorded as victims of a meteorite shower.

The distinguished British astronomer Henry Joy Stracken of Durham University, involved with researching the incident, warned that they would not be the last.[12] Stracken was closer to the truth than he could have imagined, as the dead children of Bogotá might well have joined several thousand other dead, on the boat across the Styx, if not for a quirk of fate.

[12] Sears, N, 'Death From Space: Four Children Killed as Meteorite Sets Their Home Ablaze', *Daily Mail*, Wednesday 17 December 1997.

On 9 December, five days before the deaths in Colombia, a great flash of light interrupted almost perpetual night with an artificial dawn over Greenland. Reported by three fishermen trawling Greenland's east coast, and caught on a car park's security camera in the capital, Nuuk, the flash heralded the arrival of a large meteorite (five tonnes). Bjoern Franck Joergensen of the Tycho Brahe Planetarium in Copenhagen believes it was a single meteor that crashed into Greenland's ice cap at 12,000 kilometres per hour. Greenland is such a vast territory and the weather so unpredictable that the location of the meteorite is likely to remain a mystery. The meteorite penetrated our atmosphere without detection; thousands might have died if it had had a slightly different trajectory and destroyed Nuuk. What would the consequences of our inaction have been then? Not the tragic deaths of four children, but thousands. Would we have acted then?

As absurd as it seems, politics is a beast that feeds exclusively on perceived need. Whatever motivates the electorate motivates the agents of politics. Outrage and shock at the loss of life in Nuuk may have resulted in the implementation of the EU's declaration. Politicians are unlikely to act until we, the people, demand action or until a disaster necessitates a reaction. Sadly, politics is largely a reactive enterprise and the likelihood is therefore that thousands will have to die before European governments direct any resources to this problem.

This is a matter of perception once again. If politicians perceive that there is a problem, then they will act. Jasper Wall, director of Britain's Royal Greenwich Observatory in Cambridge, feels that the problem is one of the inaccessibility of space. If we could see the world as Gagarin saw it, small and fragile but very much a part of this violent solar system of ours, then perhaps we could all begin to understand the impact risk. He tried to put the impact threat in perspective:

Look at the moon. Why do you think it looks the way it looks? Just stand out in your garden on a clear night, and look at the moon's surface. Those dark marks across its surface are impact craters. It's being battered to bits. It's being smacked into. That's why it looks like a cratered mess. The reason the Earth doesn't look like that is because it's being washed all the time by a great ball of slush.

I believe the risk is grossly underestimated. A 1 in 20,000 chance of dying from an impact is really calculated from the product of a direct hit – an impact winter of the kind that killed the dinosaurs. If you look at just how fragile our economic structure is, it is just incredible. We only need to have a small city wiped out and the economies of the Western and Eastern world would go berserk. Just a small amount of instability would wreak economic havoc.

Suppose it was determined that the impact was going to occur in Belgium, or it was going to land in a Gulf state, within a four-year deadline. Can you imagine the economic and social chaos that would result? The panic, the mass movement of people. Each country would immediately become selfish and preserve its own population. The armed forces would be mobilised. The country that was going to be hit would make damn sure that it moved, aggressively if need be, into a neighbouring territory.

Something like that might bring home how vulnerable this little planet really is.[13]

In many ways, if Nuuk had been hit our whole perspective might have changed in the fashion outlined by Jasper Wall. Are we really faced with such an unmovable and unresponsive political executive that it would take the destruction of a major city – which means homes, people and lives – simply to stir them into action? Logic dictates that we, the electorate, must use our enfranchisement to raise this issue. However, modern electoral systems are extremely blunt tools not designed to deal with specific issues. A modern technological version of the Ancient Greek system – one person, one vote, for every major issue – might be possible one day. Perhaps we could vote using an interactive digital television set. Future political maybes aside, there are few tools in the individual's arsenal that can reach and influence the political core.

Although issue-specific voting in the form of referenda has grown in popularity during the 1990s, they are usually called only to settle sovereign issues, such as taking part in European monetary union (EMU). No, traditional lobbying, grass-roots activism, stimulating of interest among environmental groups or, most importantly, encouraging the people to call for action can be the only way to begin the process of political action. Such a feat is no easy task in itself, as I will discuss later. Only hard facts, economic predictions and the possibility of making political capital are likely to stimulate

[13] Wall, J, interview, September 1998.

action by government. After all, it took nearly a third of a century of intense lobbying and petitioning by pressure groups, following the publication of Rachel Carson's breakthrough environmentalist work *Silent Spring*[14] in 1962 (reissued in 1982), to secure a place for green issues on the world political agenda, culminating in the 1992 Earth Summit in Brazil.

Our world political system has developed to respond to pressure in this way, but it is desperately slow and cumbersome. The problem facing humanity is one of time as our tomorrows rapidly become yesterdays. We must politicise this issue immediately – a subcritical object could explode over Seoul or Delhi tomorrow. Not in 50 years, or even 100,000 years. The problem is potentially immediate and, as Jasper Wall said, the risk is still little understood and grossly underestimated.

One problem that weighs heavily on the minds of all who study this threat is the question of whether our governments are truly inactive and inept, or whether they have some hidden motive for their silence. Ask yourself which scenario is worse: a government that deals in international whispers while leaving its people to live in ignorance and fear, or one that is inactive and risks the lives of the nation it is meant to protect?

[14] Carson, R, *Silent Spring*, Hamish Hamilton, 1962.

FIVE

LIES, DAMNED LIES AND CONSPIRACY

It is within the remit of every politician to obfuscate and waffle, to hide behind non sequiturs and metaphor. They have good reason to behave in this manner: they hold each state's national security in their hands. Defence, budgetary and civil matters must be handled with diplomacy and a certain amount of subterfuge. Nevertheless, it is clear that our governments are content to hide information about problems and technologies that could have a very important part to play in our lives.

Ask yourself this question: if technologies have been developed to protect Western or, more likely, American airspace from nuclear attack – via intercontinental ballistic missiles (capable of delivering a warhead from one continent to another) – would you want to know about it? It is evident that the Western governments have ground-based devices – they are well known – but what of more advanced defensive weapons that operate in a nonterrestrial theatre: space? Would you want to know about a so-called Star Wars system such as Ronald Reagan's Strategic Defense Initiative, or SDI?

Efforts to ameliorate the impact threat can be broken down into three areas: detection, determining a line of action, and taking defensive measures. If a custom detection system, a new decision-making process and a defence system need to be built and tested every time a major NEO hazard appears, our society may be unable to respond to the threat in time. If, however, such a Star Wars initiative exists, then we may have technologies in place that can be altered, customised, in order to fulfil the detect-decide-kill criteria needed to counter the asteroid, meteoroid or cometary threat. As we shall see, circumstantial evidence is growing that there is an operational Star Wars technology, operating from terrestrial bases and two tiers of satellites in orbit around Earth. The evidence is striking, yet we – the common people – know little of its existence. Why? Strategically, it is obvious that making a secret ring of defensive technologies common knowledge is ludicrous. More importantly, any missile-based orbital weapons platforms would be illegal, in direct contravention of a number of treaties, including the 1972 Antiballistic Missile Treaty (part of the Strategic Arms Limitations Talks Agreement of the same year) and the 1967 International Concord designed to preserve near-Earth space as a nonmilitarised zone.

It is just possible that the laser weapons technology first proposed in 1983 by President Reagan as SDI is now in operation. It is unlikely that the US would spend $20–30 billion developing detection technologies, which irrefutably exist – such as those operated by the US Air Force Space Command – without developing corresponding methodologies for the destruction of nuclear missiles. Certainly, aircraft can shoot down missiles; surface-to-air and sea-to-air antiballistic-missile missiles can also carry out this function. Yet, despite the limitations imposed on signatory nations by the Agreement Governing the Activities of States on the Moon and Other Celestial Bodies (1967), Article 3 of which categorically states that no weapons of mass destruction should be placed in orbit, terrestrial technologies were deemed insufficient in the 1980s and the SDI was proposed as a far better defence system. If Reagan could talk freely of the SDI in the 1980s, how far has the technology really developed since then? And can it be deployed or adapted to protect Earth from the impact threat? Most important of all, will the world's governments tell us of an impact threat, before and during the event, especially if these technologies were to be deployed?

Questions, questions . . . The frustration of ignorance coupled with the fear that we may well be kept in the dark as to the truth of events unfolding beyond our atmosphere, drives one to find answers. There are two lines of argument about the apparent secrecy surrounding detection and defence systems: either they exist and the governments are operating a conspiracy of silence, or our nation-state governments have been inactive and foolish, have failed in their responsibilities to protect national security. If the latter is true, our governments must be seen as both ignorant and foolish or simply negligent. If the former, then at least there is a reassurance that some form of technology exists to combat nuclear threats that might in turn be redeployed to counter the impact threat.

Inferences can be drawn based upon answers and hints found within the weighty pages of the NATO handbook, reports by the US Defense University and Senate Authorization Acts for the US intelligence community's budgets (*Intelligence Authorisation Act for the Fiscal Year 1996*, 29 September 1995). Furthermore, reports prepared by pressure groups such as CND, newspapers, anecdotal evidence originating in areas around restricted air-force bases and information released to military hardware manufacturers all paint a strong circumstantial picture of the hidden reality. Despite the Clinton administration's condemnation of the SDI in 1993, it seems inevitable that the technologies developed have been deployed. It is a matter of history that some of the guidance and detection software developed during the SDI research failed to enable Patriot missiles to intercept many of Saddam Hussein's Scud missiles during the Gulf War's Operation Desert Storm – proving not only that some SDI by-products were ineffective, but more crucially that the technologies reached practical deployment stage (albeit in a more terrestrial theatre).

I do not believe motives behind the secrecy are necessarily sinister, nor do I believe the Western military are conniving and driven by motives of exercising hidden power over the masses; nor are they stupid. That is simply ridiculous, or at least too frightening a prospect to be considered. No, I believe that the soldiering sector is invariably motivated by the urge to protect and defend – a cause not without merit. Nor do I condemn governments for exercising caution where matters of strategic importance exist, but I do call for more honesty, a trust in the people of the nation-states. We can and should be privy to military secrets. I am not asking to be privy to the specific processes involved in producing a new weapon or the physical characteristics of a satellite or a detection system. But I would expect the world's governments to inform us if there was a serious threat to our civilisation, or an existing means of deflecting it.

Unfortunately, few governments have good records when it comes to telling the public the truth, least of all Britain's. The well-known BSE debacle which saw the then British Agriculture Secretary, Douglas Hogg, stating that there was no risk at all to the consumer from eating British beef was recently revealed to be an interpretation of the truth so as 'not to alarm the public'. The government's own advisers had warned at the time that there was potential danger of contracting CJD, the human form of BSE, but despite this warning the government's declaration was to the contrary. Its record on such matters is appalling.

Worse still were the wrongs committed, and largely maintained to this day, against members of the armed forces immediately following the first nuclear detonations in 1945. Many experiments were conducted, such as at Christmas Island, during the 1950s, which saw troops exposed to radioactive fallout wearing little more than combat fatigues and goggles. I have a very personal connection with this sort of military lie – my grandfather visited both Hiroshima and Nagasaki on a goodwill visit. He was a chief petty officer aboard HMS *Bermuda*, a Fiji class light cruiser. *Bermuda* was a big ship – 900 crew members in wartime, 10,500 tonnes (full load), capable of 31.5 knots and bedecked with three triple six-inch guns.[1] *Bermuda* was sent along with the Pacific Fleet as a conciliatory gesture to the smashed Japanese cities, carrying supplies. The crewmen were encouraged to visit the cities, only a few months after the uranium and plutonium bombs were detonated. My grandfather inevitably died of cancer. Many of his shipmates fared little better, though some survived their battle with that particularly horrible tool from the Reaper's arsenal.

I remember he spoke about the horrors he and his colleagues saw: a city in ruins, none of the debris cleared, charred wood, twisted metal and scorched earth. He was mesmerised by the rivers of glass that had been blown from windows and glassware in the Japanese homes; they must have melted in the intense heat of the fireball and later solidified into amorphous paths of glass. The glass was one of the few things that were recognisable in the wreckage. The crewmen walked through this debris, through the radioactive fallout that still covered the ground. Although they had been told not to take anything back as a memento, they were not warned of the radioactive dangers inherent in such a site. Grandfather's cautionary tale always ended with his express wish that the people who had authorised the use of the A-bombs could have 'experienced' the desolation first-hand: the burnt children in hospital wards – screaming in agony while their dressings were changed – dying slowly after the blasts. If the decision makers could have smelt the children's wounds and the general reek of sickness, he felt sure that they would have been forever haunted by their guilt.

The terrible damage inflicted upon the Japanese by the atomic blasts aside, the crewmen of HMS *Bermuda* were just some of the many troops exposed to radiation and then left in ignorance to

[1] Roberts, J, and Rowe, A, *British Cruisers of World War II*, Arms and Armour Press, 1980.

deal with the consequences in the decades that followed. Many ex-servicemen, particularly in the USA, have attempted to seek retribution and compensation from their own governments – it is a hard and often fruitless battle. I say again: few governments have good records when it comes to telling their citizens the truth.

I fear that if Earth were threatened by an impact, particularly one that might occur with only a few days' or perhaps hours' notice, we would not be told. Frankly, there would be little point – what are governments to do? They could not prepare a defence in a few days. Not unless there was a specifically designed NEO defensive system already in place – but currently there is no such system in existence. We would, I suspect, be left to die in ignorance. After all, what would we do if we were told that our city was to be the target impact site? Would we block roads and transport systems as we tried to leave the area or country? Which nation would allow us entry in any case? I can think of none that would be willing to allow an influx of millions of foreign citizens in one day. This attitude of telling the public 'only what they need to know' is almost certainly one of the key motivations for playing down the importance of impacts.

Despite the attempts of numerous scientists, the father of the hydrogen bomb, Edward Teller, included, the world's nation-state governments refuse to move the issue to centre stage. It seems that only an upsurge of opinion like that seen over the much less ominous effects of global warming will secure a political action. Currently, concerns about impacts operate at a much higher level and attempts to bluff and obfuscate by the military complexes around the world have failed. The British Ministry of Defence has claimed on many occasions to have no involvement in studying impacts and yet it was the MOD that responded to Edward Teller's letter to John Major in 1996. The MOD said:

> As you may know, the MOD has been made aware of the concerns being expressed as to the threat from Near Earth Objects – primarily as the result of approaches from Major Jonathan Tate, with whose work you will no doubt be familiar. Although recognising the concern, we do not believe that, in the context of competing priorities for defence resources, funds should be made available from the defence budget to support this work.

This trite response was made despite the fact that many space scientists, world geographers and anthropologists and even NASA (America's own space agency) believe the problem is of the highest order. The impact threat is obvious; it seems unlikely that the British or American military would ignore such a threat to their realms. There is a political subterfuge, as evidenced by the attempt to minimise fears over the JA1 asteroid in 1996. Will democratic society be allowed to come to terms with imminent collisions in the future? Unfortunately, history suggests not.

Ever since Orson Welles's famous radio broadcast of *War of the Worlds* in 1938 – which caused mass hysteria by appearing to be real – the US government has been paranoid about the announcement of impending disaster. Perhaps they have just cause, as in 1910 mass hysteria caused many suicides in Denver, New York, Pittsburgh and San Francisco as people became depressed over the apparent 'end of the world' when Halley's Comet reappeared in our skies. Certainly, my own grandmother, born of a more Victorian generation, was similarly fearful of the comet.

Despite such events in the past, is it not time to re-examine this secretive philosophy? We are not children, nor is the populous on the whole ignorant. I feel that we are now more than capable, trained as we are by daily news broadcasts, to understand and rationally assess well-worded warnings from our government. Evidently, as we have seen in the case of BSE and other recent public-relations farragos, our governments do not agree. There is and always has been a conspiracy of silence where matters of safety to the mass populous arise. Seldom has there been honesty; the distribution of gas masks on the home front prior to World War Two[2] was one of the few exceptions to this wall of silence. Even then, the Allies told the people that they faced potential gas attacks only because they feared that the engines of industry – the workforce – would be severely depleted, effectively ending the war effort. The workforce had to be told if the war were to be won with weapons made in the factories of the home front.

[2] The British parliament announced that children and families would be given gas masks in January 1938. This was in part a reaction to memories of chlorine gas attacks in the trenches of World War One, the gathering momentum of war in Europe and Mussolini's use of poison gas during his efforts to annex Ethiopia in 1935. This occurred despite the creation of an international convention outlawing the use of poison gas in 1925.

With these thoughts in mind, is it not possible to encourage the light of reason to be lit in the minds of politicians if the dangers of impact are presented in a way analogous to the risk assessments done during the interwar period? That said, I still do not believe, perhaps cannot face the consequences of believing, the idea that our governments have been entirely inactive in developing relevant technologies. Granted, they are not specifically designed to destroy would-be impactors, but there is technology nonetheless that could be modified to perform that deflection function.

In 1998, the British and American governments revealed intelligence that strongly intimated that Saddam Hussein's Iraqi government had prepared biochemical weapons of mass destruction – such as anthrax – which could be sprayed from specially adapted pilotless aeroplanes. Why did they tell us this? Why hadn't they revealed these fears prior to Desert Storm in 1990? It is evident that the Allied governments felt they needed to justify their December 1998 bombing campaign, known as Desert Fox, to the world. Would they have told the public if they could have destroyed the weapons by stealthier, more covert means? I think not. Nations afraid of what is a very real threat from Iraq will tolerate high-profile bombing raids. Yet without the revelation about the existence of these secret Iraqi weapons, who would have supported President Bill Clinton and the British Prime Minister, Tony Blair, in their cruise-missile-led attacks? The Iraqi leader had to be stopped, so they revealed their justification. I am certain that no such revelation will be forthcoming about the suspicion in journalistic circles that SDI technologies have been used to detect and destroy missile installations. Some of Iraq's infamous Scud missiles, referred to by the press as 'flying dustbins', failed to meet their Kuwaiti and Allied targets during the Gulf War. Why? Are the Iraqis and their former Soviet suppliers really incapable of preparing and launching effective missile strikes? Unlikely.

It has become apparent, to those interested enough to monitor budgetary proposals and the development of installations around the world, that SDI-type detection technology was used to spot and negate Scuds at launch, in and after the Gulf War and especially during the more recent Desert Fox operation in 1998. There is a feeling among many of my colleagues in the British press (a much maligned breed though it be!) that the SDI, little spoken of since the late 1980s, is now fully operational. It is thought to be much more classified than the F-117A Stealth fighter and its bigger brother, the

B-2 Stealth bomber, or other advanced technologies – such as smart bombs (Sensor Fused Munitions for Artillery Ammunition, Explosives and Pyrotechnics[3]) and SADARMS (Sense and Destroy Armor Munitions) – now commonly used in the theatre of battle.

The system operates in a number of ways: a variety of detection methods all orchestrated by one central command known as US Space Command. There is a terrestrial system of electronic optical telescopes known as the Ground-Based Electro-Optical Deep-Space Surveillance System (GEODSS), along with the Space Detection and Tracking System (SPADATS), which includes the US Navy's Space Surveillance System, the US Air Force's global network of cameras and radar (known as Spacetrack) and data from Canada's satellite-tracking network. This series of detectors is further enhanced by the Ballistic Missile Early Warning System (BMEWS) in orbit round Earth, the Pave Paws submarine-launched missile-detection network ('Pave' is the name of a US Air Force programme, while 'Paws' is an acronym for Phased Array Warning System) and the North American Defense Command (NORAD), which tracks and records the movements of all satellites, giving warnings of orbit deviation to the air force's control facility. The US Space Command, built beneath millions of tonnes of rock in Cheyenne Mountain, Colorado, is quite simply the hub of the most comprehensive surveillance and detection system ever devised. Bear in mind that in addition to these systems there are also a number of strategic 'spy' satellites in orbit which can be retasked as 'wild cards' in order to detect and track specific groups (Iraqi weapons caches, for example) or target specific locales.

Rumours from a military base resulted in reports by the BBC[4] which claimed that RAF Menwith Hill, near Harrogate, North Yorkshire, is a base dedicated to facilities that feed the USA with covert information and employs the orbital Space-Based Infrared System (SBIRS). The reports went further, stating that the SBIRS detects the ignition of missile engines and conveys that targeting information to orbital weapons platforms which then destroy the enemy missile

---

[3] Smart bombs are now used extensively (exclusively during the Desert Fox engagement) when accuracy of bombing and minimisation of civilian casualties is paramount.

[4] *Close Up North*, BBC Television, 4 December 1998, and 'UK Base Is Star Wars Station and Saved Thousands of Lives', BBC News Online, 5 December 1998.

target from space – probably using a laser weapon. The BBC report was quite specific in its allegations, stating that the strangely shaped spherical buildings have played active roles in the various engagements in the Gulf in the 1990s, perhaps saving thousands of lives by detecting Iraqi missiles at launch and neutralising them. It is evident that no weapons exist at the UK base, but certainly it appears that detectors do operate from within the Golf Balls, as we locals (or 'ex-locals' in my case) have always called them. Even when I was a child I used to speculate about the nature of the Golf Balls – we always thought they had some major part to play in defence but never actually knew what.

The base is managed by the US National Security Agency and certainly operates as a major 'listening post' monitoring electronic and voice transmissions for information vital to national security. The site covers 560 acres and is also home to the 713 Military Intelligence Group and the 451 Intelligence Squadron.[5]

The SBIRS is broken down into two tiers – the high and the low. The high phase is said to give vital information about enemy missile-launch events, while the low phase (a ring of 24 satellites) is thought to enable the detection system to plot accurate impact predictions and determine trajectory.[6] The central hub for the SBIRS is thought to be in Denver, Colorado, at Buckley Air Force Base, where another set of 'golf-ball' protective domes (constructed of a plasticised fabric akin to that which covers the Millennium Dome in the UK) have been erected, as at Menwith Hill. Estimates place the cost of the SBIRS system in the region of $12–20 billion. The Menwith Hill base falls under the auspices of the CIA. Funding issues, even for matters as mundane as creating better childcare/crèche facilities, have to be authorised by the CIA and/or the Senate, as this extract from the US intelligence agencies' budgetary bill, 1996, demonstrates:

SEC. 504. ENHANCEMENT OF CAPABILITIES OF CERTAIN INTELLIGENCE STATIONS.

(a) Authority: (1) In addition to funds otherwise available for such, the Secretary of the Army is authorized to transfer or reprogram for

[5] Pike, J, FAS:RSOC Menwith Hill, UK, www.fas.org/irp/facility/menwith.htm.
[6] Olgeirson, I, 'Buckley Revamps Missile Watch', *Denver Business Journal*, 20 October 1997.

the enhancement of the capabilities of the Bad Aibling Station and the
Menwith Hill Station, including improvements of facility infrastructure
and quality of life programs at both installations.[7]

The fact that the CIA is, strictly speaking, responsible for the
budgetary requirements of Menwith Hill Station – even though
the above extract focuses on accommodation as opposed to equipment installation – proves that the base has an intelligence-gathering
function. The similarity between the form of geodesic protective
plastic domes in Denver and those in Yorkshire (a common form of
protection which has also been employed by Britain's Defence
Research Agency to cover expensive satellite dishes) and the fact that
the base is suspected by numerous groups – such as the BBC and
Greenpeace – to play a part in the new SBIRS keenly indicate that
the Menwith Hill, Buckley (Denver) and even the US base in Thule,
Greenland (which has been an operational base for the BMEWS
system[8] and may well have been upgraded too), all have a part to
play in the implementation of a vast, new detection system of the
kind outlined in President Reagan's initial Star Wars speech in 1983.

No organisation would, I believe, invest in such a complex array
of detection systems without having a similar number of advanced
defensive options, some of which must logically be space-based. It is
certain that ground-launched antiballistic-missile systems exist, but
clearly an orbital launch system would offer a number of tactical
advantages over these more conventional systems. If not, why
propose the Strategic Defense Initiative in the first place?

The value of these incumbent detection systems to the safety and
maintenance of Western security is incalculable. The technology is
highly innovative and could possibly be used, as in the ground-
based NEAT (Near-Earth Asteroid Tracking) programme, to detect
naturally occurring stealthy objects such as potentially hazardous
comets, asteroids and meteoroids. But this technology could be reconfigured in this way only if it became redundant. The Iraqi leader,
Saddam Hussein, may continue to threaten the West for some time,
as might the Libyan leader Colonel Gaddafi, and Islamic fundamen-

[7] Intelligence Authorization Act for the Fiscal Year 1996, Senate, 29 September 1995.
[8] Williams, C, Tech. Sgt, 'Serving at the Top of the World', Air Force Link
News, http://www.af.mil/news/, 1998.

talists such as Osama bin Laden. While their threats and those of other nations persist, detection systems will be needed; but what if the orbital systems were superseded by ground-based systems? There is a strong possibility that just such a terrestrial system may be operational in the short term. It depends upon the manipulation of our atmosphere.

The sun attacks our planet's atmosphere, spewing out millions of tonnes of deadly radioactive material on a daily basis. Protecting us from this constant bombardment is a shield of electromagnetic waves created by Earth's magnetic field: the magnetosphere. This ongoing battle is often revealed when our magnetosphere falters, allowing trails of charged particles to spiral into our atmosphere, streaking the sky with ribbons of colour. Known as the aurora borealis, or northern lights, these strips of radioactive material can have severe effects on electronics and have even been known to corrode Alaskan gas pipelines. Other particles, released from the solar matter, are trapped within the magnetosphere, creating the ionosphere (55–800 kilometres above our planet's surface). A US military research project, built near Gakona on the rolling tundra of polar Alaska, is set to turn the ionosphere into a mechanism of covert war.

Managed by the US Navy and Air Force, the High-Frequency Active Auroral Research Program (HAARP) is designed to give the US a greater understanding of the aurora effects and to stimulate in a limited way similar artificially created effects so as to develop a greater understanding of our ionosphere. Few would be coy enough to believe such an asinine description. It is most likely that HAARP could disrupt communications, destroy electronics, even destroy missiles, with directed radio waves in much the same way that an EM (electromagnetic) burst from a nuclear weapon would. Furthermore, HAARP could achieve this in a remote fashion from its Alaskan base.

Construction began in 1993; testing of the initial installation began in 1995. Success soon followed. The massive antenna array, comprising 180 thirty-foot-high transmitters, blasted energy up beyond the stratosphere into the ionosphere. The ionosphere was heated, and produced a precisely focused beam of extremely low frequency (ELF) waves, which could penetrate hundreds of feet below ground level. By measuring both the time these signals took to return to the antenna array, plus the quality of the returned signal, an image of a disused mine was produced. HAARP has proved to be so successful that funding has been secured to change this most powerful of experimental projects into a fully operational prototype by the year 2002.[9] HAARP will be able to transmit ten times the power that the current experimental project can achieve: reaching 36,000 kilowatts within five years.

The Pentagon's prototype is based on decades of research into the solar winds. Victorian scientists noted that turbulence on the sun's surface creates repulsive forces of immense magnitude. These forces eject matter from the sun, out into the solar system. Our magnetosphere is buffeted by this so-called solar wind. Ultraviolet light and X-rays penetrate our planet's shield. They impact with our oxygen-rich atmosphere and create a field of charged particles, or plasma, round the Earth. This is the ionosphere.

Many have tried to manipulate its power, but the unwitting father of ionospheric warfare was the Croatian-born inventor Nikola Tesla (1856–1943). Known as the Magician of Wardencliffe, having emigrated from Croatia to New York in 1884, Tesla was a true genius. Renowned as the man who first illuminated the world, he was the inventor of alternating current, power lines and electrical grid systems, and lodged over 700 other technological patents – many of which are still classified, for military eyes only.

Tesla frequently fell foul of his contemporaries. His seemingly outlandish vision of an intricately computerised society, dependent on long-distance wireless communication, was more than the scientific community could bear in the early twentieth century.

Tesla began a long-term experiment to harness the ionosphere in 1901. It was an element of our atmosphere that had not even been discovered by orthodox science at that time, yet Tesla not only determined its existence but also set out to exploit it. He believed in the existence of an electroconductive sheath in the atmosphere, capable of carrying huge quantities of power around the world, without the need for wires. Tesla had visions of illuminating the Paris Exposition with hydroelectric power generated at Niagara Falls and transmitted to Paris via the ionosphere. The Magician of Wardencliffe also hoped to use the ionosphere to broadcast around the world, bouncing his signal off the ionosphere as if it were a mirror in space. This would enable the signal to travel around the curved surface of the globe, as radio signals effectively radiate in straight lines,

[9] http://w3.nrl.navy.mil/haarp.html.

behaving in much the same way as the visible part of the electromagnetic spectrum, light.

He failed. Marconi beat him to the goal of transmitting across the oceans, without the need for a giant tower. Marconi also employed the ionosphere as a 'mirror' to bounce his radio signals around the curvature of the Earth.

In the 1950s the US military began to think again about Tesla's theories. They initiated an intensive study of the ionosphere, in an attempt to control it for military purposes. The military complex's interest was piqued when nuclear-weapons tests carried out at the edge of the atmosphere revealed the disruptive effects of electromagnetic pulses released by nuclear explosions at detonation. Vast artificial auroras were produced, affecting electronics and electricity supplies over the Pacific basin, particularly in Hawaii. It became apparent that an electromagnetic pulse could comprise a tactical weapon in itself; the effects of radiation were minimised because the bombs were detonated outside the life-supporting zone of the atmosphere. The atmosphere itself acted as a barrier to a vast majority of the radioactive fallout emitted at detonation, allowing only plasma and electromagnetic radiation to flood the atmosphere, which undermined radio and television broadcasts, electrical devices and power supplies in the process.

It was far from a precise art and many international scientists, including Sir Brian Lovell, Britain's leading astronomer, condemned the tests as foolhardy. Very real risks had existed that one of these great hydrogen bombs would explode prematurely, flooding the lower atmosphere with deadly fallout (radioactive dust) or cause some other unforeseen consequence. The US military was left with a tantalising question: could this electromagnetic (EM) pulse be simulated or produced in another, controlled, way, over enemy territory, without the consequential long-term contamination and collateral damage caused by nuclear explosions and fallout?

This question, combined with new rocket technology – born of culled Nazi V2 research – led to the discovery of what became known as the Van Allen belts: layers of radioactive belts on the upper fringes of our atmosphere. The US wanted to use an EM pulse as a weapon and, as we shall see later in the chapter, new ionospheric research may give them that ability. More importantly, this new research may make other space-based technologies redundant, allowing them to be reprogrammed and used to protect Earth from the impact threat.

In October 1962, the Soviets tried to locate nuclear missiles on the island of Cuba, just off the southeastern shore of the United States. Few could ignore the potential link between aggressive US atmospheric nuclear tests and the Soviet bloc's attempts to deploy nuclear missiles on the fringe of US airspace.

A peaceful resolution was reached to what became known as the Cuban Missile Crisis, after a protracted period of friction, barely avoiding a third world war. The Limited Arms Test-Ban Treaty swiftly followed it, in 1963. The US Defense Department's drive to gain control of the ionosphere had been effectively terminated.

Twenty years later, they were to be given another chance – and this time there would be no turning back. During the development of the SDI, a large auroral research programme was initiated with the aim of manipulating Tesla's upper-atmosphere energy belt (the ionosphere) in order to generate powerful forms of low-frequency waves. It was hoped that this would lead to the use of waves as a means of destroying intercontinental ballistic missiles as well as detecting underground buildings and complexes, and perhaps even rocket sites. These very low-frequency waves could, in essence, affect electronics and other modern technology in much the same way as EM pulses generated by atmospheric nuclear explosions. By controlling how much energy is pumped into the ionosphere and directing that energy, it is possible to use these low-frequency waves in much the same way that radar and sonar use directed energy.

Scientists developed a range of technologies for manipulating the ionosphere. Pivotal in this research was the ionospheric heater. This machine could transmit a beam of radiation directly at the ionosphere, to heat up an area of the plasma field. The electromagnetic wave hit the plasma and caused electrons and neutral particles to move, which in turn caused friction and heat. Just as a microwave oven functions by vibrating water molecules to generate heat within food, so does an ionospheric heater with plasma.

The development and use of ionospheric heaters, with the SDI, coincided with a problem on the gas fields of Alaska. Arco (Atlantic Richfield Company) had billions of cubic metres of excess gas to dispose of. It had proved to be commercially unfeasible to build direct pipelines to the US. They were at a loss as to how to dispose of such enormous excess. They turned to Tesla's old concepts for manipulating the ionosphere, using the surplus gas to power an antenna array capable of making enormous quantities of radio waves.

Alaska is located close to the Earth's magnetic pole. If they could transmit masses of radio waves from ionospheric heaters and hit the upper regions of the magnetic field, the US could make the ionosphere produce a shower of electrons at 50 billion degrees Celsius. The military also knew that a likely trajectory for Soviet ballistic missiles would be a polar one.

The US military complex now had a concept that could change the path of global warfare. This was a new kind of warfare, conducted by remote means, in the invisible world of electromagnetic bombardment. The HAARP team designed an antenna array, not unlike Tesla's tower at Wardencliffe, but on a much larger scale. The antenna would bombard the Van Allen belts with particles and build a shield of electrons over the United States.

The US Defense Department insists that the system's main function is to monitor weather patterns and to attempt to constrain the effects of the aurora borealis on satellite communication and electronic equipment analogous to the effects of EM pulses. It claims that the ability to monitor the construction of enemy underground weapons bunkers is merely a by-product. Yet it is now well known that one of the first experiments conducted with HAARP was to image (detect) a disused mine near Fairbanks, Alaska, a town near the HAARP array. It seems that HAARP's ability to operate as a covert 'tool' was appreciated as early as its first month of testing. My own journalistic instinct suggests that those members of the military who are bank-rolling the HAARP will rightly expect a good deal more from it than a very accurate weather report.

It is impossible to predict whether the operational upgrade for HAARP will enable it to exercise true control of the ionospheric environment, or even if the US Defense Department intends to try to exercise such control. What is clear is that the initial tests where HAARP was used as a detector produced striking results.

It is conceivable, then, that the military orbital-satellite-based detection systems might be made redundant by the HAARP project if it proves an effective detector of missiles. If so, then there will be a great deal of highly advanced detection equipment in orbit which has the potential to be reconfigured and used to detect potentially hazardous near-Earth objects. Whereas the detection systems were designed to track hot signals from weapons, it is not inconceivable that they could detect cold, yet similarly stealthy, natural objects of destruction. It depends upon the actual use, configuration and success rate of the HAARP. I hope that it succeeds as a detection system and that the US can be convinced that its expensive orbital hardware should be redeployed to look at the solar system to track incursions into near-Earth space.

It is a reflection of the seriousness with which the US military now consider the impact threat that Air Force Space Command is considering the possibility of using its sensors to detect NEOs. Indeed the ground-breaking NEAT programme was established under the guidance of NASA's Pasadena-based Jet Propulsion Laboratory, which set out to do just that, exploiting the GEODSS site at Haleakala on the Hawaiian island of Maui. Initiated in December 1995, the NEAT project allows astronomers access to the system six nights per month. The GEODSS depends upon a triumvirate of telescopes – one primary one-metre scope and two additional ones, each equipped with extremely sensitive 4,096-by-4,096-pixel CCDs (charged-couple devices). These are like those found in a domestic video camera but of much greater resolution – able to detect space debris (such as that from shattered satellites or decaying fragments of launch vehicles) or NEOs at light levels 10,000 times lower than that possible with the naked eye. It is a remarkable system. The three main sites at Haleakala on Maui, Socorro in New Mexico, and Diego Marcia in the Indian Ocean, all report to the 21st Space Wing, headquartered at Peterson Air Force Base, Colorado, though the NEAT programme is managed by the Jet Propulsion Laboratory in Pasadena.

Despite the limited 'lens-time' allowed to astronomers for NEAT on the Maui telescopes, an enormous amount of data is received that demands huge computing power – thanks at least in part to its excellent location 3,000 metres above sea level in largely clear skies. A new computer system was installed in May 1998, effectively enabling the investigators to double their coverage of the heavens. This new analysis hardware and computer system was installed and funded by NASA under its new NEO initiative, which has made an extra $3 million available for NEO detection this year.

The system includes four 300-megahertz computer processors, each of which is entirely dedicated to the NEAT data. One of NEAT's principal investigators and project manager, Dr Steven Pravdo, was optimistic that the new equipment would produce remarkable results:

This new system will speed up the processing of data and allow us to analyze up to 40 gigabytes of data each night, or the equivalent of nearly 70 CD-ROMs. We will be able to double the amount of sky we search each night, which is currently 500 square degrees, as well as the number of new asteroids and comets we find during each monthly observation cycle.[10]

Pravdo's optimism was soon proven well placed. Having already detected approximately 700 asteroids by January 1997, the new computer equipment enabled the NEAT team to detect two new large Earth-crossing objects within a month of installation. The new objects – asteroids 1998 OH and 1998 OR2 – were categorised by the Minor Planet Center (Harvard Smithsonian Center for Astrophysics, Cambridge, Massachussets) as potentially hazardous asteroids (PHAs). These are two of 125 such objects currently catalogued (see Appendix). Each of the new discoveries exceeds the one-kilometre size (approximately 1–3 kilometres) generally assumed to be the watershed for generating globally devastating effects and mass extinctions if they were to impact with Earth. Although neither of these objects is likely to pose a threat in the next few decades, these new discoveries prove the value of military and civilian symbiosis.

I believe sovereign states around the world ought to use this highly successful model as a basis for their programmes, each developing a catalogue of these objects, and more importantly developing the scientific knowledge and techniques to detect such things. If a more persuasive reason is needed, it must be obvious that the technologies and skills needed to track NEOs is exactly the same as that needed to detect foreign spy satellites, missiles and other objects, including space debris. Surely that is a valuable resource for any nation.

There is a worrying tactical element to the likely wall of silence that might precede impact: there is a very real threat from misunderstood footprints of exploding asteroidal, meteoroidal and cometary material. The United States military complex has, as I have outlined, a staggering web of detectors both terrestrially and in orbit, yet even this most sophisticated of spiders was reputedly fooled by explosions: the US was put on full alert by 'nuclear explosions' over the Pacific in 1994.

Imagine the scenario. It was evening in the USA. Spacecom was

[10] 'Latest Computers Will Boost Asteroid-Tracking Efforts', NASA Jet Propulsion Lab. PR: 20 May 1998.

operating normally, its vast array of detection devices monitoring the orbits of the nearly 3,000 higher-orbit geostationary satellites. The nuclear early-warning system was at rest – thankfully no activity to report. In Washington, DC, the picture was a little different: the nation's leading politicians were fraught with tensions as they attempted to negotiate with North Korea over their declaration of intent to withdraw from the Non-Proliferation Treaty (signed internationally in 1968 and renewed in 1995), which was designed to prevent the spread of nuclear weapons. President Clinton slept.

While this dynamic was in place, a small stony object approached Earth at high speed – possibly as high as 38,000 kilometres per hour. It entered our atmosphere, air friction superheating its surface as it plummeted towards the Pacific Ocean.

The object reached critical temperatures – it could no longer withstand our atmosphere's battering – and so, 20,000 metres above sea level, the object exploded with the force of approximately five Hiroshima-sized A-bombs: 50–70 kilotons of TNT.

The US Department of Defense's detection system picked up the explosion: it had all the hallmarks of a nuclear detonation. The system went to full alert – it is rumoured in journalistic circles that Clinton was roused and asked to stand ready in case of need for retaliation or other action. Remember, this was a period when the US was engaged in heated substantive talks with Korea over its nuclear programme.

Only further analysis of the data cast doubt over a terrestrial origin. Someone interpreted the data – probably due to the lack of a launch-vehicle heat signature – as probably being caused by a highly energised impactor exploding in the atmosphere. The alert was over, but the system had apparently been fooled enough to trigger a nuclear alert.

We will probably never know the full events of that night, and in a sense we do not need to; the fact is that approximately 250 small bolides have exploded in our atmosphere at high altitudes during the 1990s. The US detection web picked up these explosions. What if a mistake was made?

More disturbingly, what if another, less well-equipped, nation misunderstands the origin of the explosion? There is a great ignorance of this problem in political circles. Ask most British politicians about this element of the impact threat (or the impact threat in general) and they will no doubt stare at you blankly. The impact phenomenon has many guises; the most common and potentially unsettling guise is that of perceived nuclear explosion.

Think of it – there are many 'trigger-happy' nations, each flexing their military muscles in an effort to outmanoeuvre rivals: Iraq, Turkey and Greece, Pakistan and India, China and Taiwan, to name but a few. The consequences of a high-altitude – or, worse still, a low-altitude – explosion of the type witnessed over Korea in 1994 might be catastrophic at the wrong moment. Indeed, there would be damnation from all quarters of NATO if it were believed that a small nuclear device had been detonated over Iraq by the US or Britain (surely the most likely suspects, bearing in mind their coalition during Desert Fox in 1998). It would be very difficult indeed to prove that such an explosion (whether it did harm or not) was triggered by a bolide. The ambivalence directed towards the UK and US following Desert Fox, even by fellow UN Security Council members such as Russia, sets the scene for a dangerous backlash if we are unlucky enough to suffer another Korea-type explosion.

That sort of eventuality is more likely than you might expect, as 250 high-altitude explosions over a decade equates with 25 per year on average, or one per fortnight. Unfortunately, that is not the only strategic problem.

A knowledge and skills gap already exists between the US and the rest of the world. We are all, to a large extent, beholden to the US for our information on this subject. Theirs is the only nation, thus far, to seriously take a grip of this problem – even then, only in a tentative way. What if the USA decides to withhold intelligence that its detection network has produced concerning an impact threat? Asteroids could be seen as a strategic weapon: the US may not withhold impact data from the UK but it may view Iraq in a different light. Should not all nations therefore have an NEO project, for national-security reasons alone?

This conspiracy of silence is reflected in the 1996 JA1 close approach: no one was told; no one even knew the object existed until it was approximately four days away from Earth, and yet when it was detected no alarm was raised by our respective governments. Remember, not even Spacecom knew if JA1 would impact until the object was approximately one day away from Earth. Yet our nation-state governments left us in ignorant bliss – Spacecom would not have been able to delineate a precise impact trajectory for JA1 until a few hours before impact. It could have destroyed a significant

portion of Continental Europe or the British Isles, or missed altogether, as it eventually did – but we were not given the opportunity
to brace ourselves. Why was the public kept in the dark? Was it that
there was no mechanism in place for the dissemination of the news of
JA1's arrival in a near-Earth trajectory? Or is it that our governments
wanted to exercise the 'need-to-know' policy so frequently employed
in the twentieth century?

Ignorance and political idiocy are still a possibility. The UK's
Natural Environment Research Council (NERC) is a government
body designed to manage research and development in the fields of
natural-disaster and environmental perturbation. Spaceguard UK, a
pressure group affiliated to the central Spaceguard lobby established
by Eugene Shoemaker and like-minded scientists following the Jupiter
collision, wrote to NERC asking for their support in fostering
research into the impact threat. Their response was staggering, giving
credence to the most disturbing of all possibilities: our governments
are dangerously ignorant of the risks.

Their reply stated:

> . . . in relation to the prediction and mitigation of the effects of aster-
> oidal and cometary impact, I have no information to forward to you.
> This area is not covered by the NERC mission. The work which
> NERC supports on environmental risks and hazards focuses on extreme
> natural events including inland flooding, storm surges, seismic events,
> storm tracks, land instability, and the environmental consequences of
> the release of genetically modified organisms.[11]

While the release of genetically modified organisms has no part to
play in this discussion, every other field of research mentioned in the
NERC response is directly relevant to the impact threat. Indeed,
the list of areas of natural perturbations supported by NERC reads
like a description of a catastrophic impact! What is an asteroidal
impact with the surface of Earth but a natural disaster? It is not a
man-made disaster or one created by little green men from Mars. I
find the lack of comprehension staggering. That is not to blame the
staff of NERC: they can respond only to their 'known' parameters.
Discussions and policy from government should direct them to

[11] Letter dated 7 October 1997, to J Tate, director, Spaceguard UK, from
RKG Paul, director, Planning and Communications, NERC.

include the concept in training and as an explicit feature of their remit. The fact remains, however, that NERC's remit does indeed cover this threat, albeit implicitly.

Jasper Wall of the Royal Greenwich Observatory (RGO) believes that there are political motivations within government, and the MOD in particular, for the suppression of the true threat and even avoidance of the issue altogether – denying reality in an effort to safeguard the status quo. If one enquires at the press office of Britain's Defence Evaluation and Research Agency (DERA) or the MOD on this subject, one is given short shrift. Wall has long fought internal battles over funding with government and quasi-government bodies. His insights into their processes are quite revealing, especially when one considers that he has recently fought and lost a battle to maintain the RGO, which is now to be mothballed and its staff – the cream of Britain's elite – scattered to the four winds.

> The reason that the MOD is silent is that they know the sorts of cash sums involved [says Wall]. Look at how much we spend on nuclear safety. People go crazy about the risk, and yet we have minimised the risk from our own nuclear power plants. Still people worry because nuclear power and radiation is something that they can understand, because it is a tangible perceived threat: nuclear power stations are visible and so they fear them. Yet the risk from nuclear meltdown is infinitesimal, dwarfed by the risk from impacts.
>
> This whole question of risk and how much we spend on it is the interesting thing. If the public had a perception that the risk of dying through impact disaster was anything like as high as it actually is, then the MOD would have to spend money on it. That is why the MOD won't touch it with a bargepole and why it has a moratorium on it. Because they know that, if they start talking about it and it gets picked up by the press and the general public, it would completely distort, in entirety, the way they are structured, the way their funding comes, what their funding is for, what they do, who runs the institution. It would just turn the MOD inside out.[12]

Can that be it? Are our governments and the institutions of government – intelligence, defence, research and environmental councils – so concerned about disruption to their status quo, that any problem, however frightening and potentially catastrophic, must be ignored?

[12] Wall, J, Royal Greenwich Observatory, Cambridge, interview, 1998

Can movement from established patterns really threaten our institutions that much? Surely they are designed to gather information, research and prepare plans that will protect our countries from danger – that is why they are vested with power. If Jasper Wall's insight into these corridors of power really is a reflection of reality, then it would seem that petty power brokering and fears of dislocation and demotion drive our primary organs of power into a visceral catatonia.

Inactivity, known as output failure in political circles, often seems to be the net result of our political systems. They work quite well when tackling long-understood problems, but present them with a new one and they falter, despite several years of extreme climatic perturbation acting like some great warning beacon.

In 1998 the people of Earth suffered 96 disasters in 55 countries.[13] Hurricanes Georges and Mitch had what the International Development Corporation described as 'catastrophic consequences': killing tens of thousands and leaving millions homeless. In Naples a hundred people died thanks to flooding, swamped and drowned in liquid mud. At Easter, Britain faced its worst floods for two decades, while in December the East Coast of America basked in temperatures of 20 degrees Celsius. Another timely reminder of nature's destructive power came in late December 1998, when a nuclear power plant declared a 'full-scale emergency when strong winds knocked out power-lines and forced the reactors to shut down.'[14] The plant in question was Ayrshire's Hunter B. According to reports it took staff five hours to reboot safety systems in order to prevent the core going critical. Despite the fact that the Nuclear Installations Inspectorate has launched an inquiry the government has made no official announcement on this issue.

Are we finally, after this most disruptive of years, going to accept that nature is the most awesome force we face – not petty dictators or violent terrorists, but nature? Perhaps if we do accept this fact, we can then encourage our governments to be more honest about the risk that we face from catastrophic events – not just the sort of terrible damage witnessed in 1998, but that which will inevitably be visited upon us by a vastly disruptive asteroid or cometary impact.

[13] Independent Television News, 31 December 1998.
[14] 'Emergnecy as Gales Hit Nuclear Plant', BBC Ceefax News, 30 December 1998.

111

Maybe then we can focus some of our extraordinary skills developed for manufacturing elaborate communications systems and weapons of mass destruction on protecting what we claim to value so much: life.

We are beginning to see the development of a symbiotic relationship between two key, formerly highly stratified socioeconomic groups: the military and scientific communities. It has always been a case of 'never the twain shall meet', as far as these two groups are concerned, for whenever they have met military leaders have been frustrated by scientific idealism – as in the case of the atom-bomb creator Oppenheimer's eventual rejection of his own device – and scientists similarly frustrated by what they often describe as 'the perversion of their research'. Nevertheless, despite this traditional rift in ideology, programmes such as NEAT reflect a growing interrelationship that ought to be fostered and encouraged. For only by exploiting the skills of both fields, and many others, will we eradicate this great threat from space. It is a beginning, but we need to go further and develop the mechanisms and tools to protect ourselves, or else we may inevitably face a situation where we mourn the passing of a great city, the death of millions, or worse.

CHAPTER

SIX

Shielding the Cradle

*In the year 1999, and seven months, from the sky will
come the great King of Terror . . .*[1]

So warned the great sixteenth-century physician and astronomer and,
latterly, prophet, Michel de Nostredame, better known as Nostradamus (1503–66). The King of Terror might be an apt description of
an enraged dictator unleashing some great war from the air. More
disturbingly, this quatrain from the prophecies may refer to the
coming of some great deathrock from the stars. Not wishing to decry
others' belief in the talents of Nostradamus, I find it difficult to
believe that his predictions are little more than the disturbed musings
of a very active imagination interpreted by translators over the centuries to instil awe and fear in his readers. Nevertheless, it is a timely
warning. If such a 'King of Terror' were to plummet towards Earth
from space, could we defend ourselves by deflecting a great mountain
of rock, ice and metal? Not if a 'deathrock' were to arrive in 1999:
we would simply die in our millions. On a more optimistic note (and
not wishing to tempt fate!), the prophecy may be false, giving us a
little time yet to plan, build and defend ourselves.

The topic of building a defensive shield around our world is the
stuff of adventure comics. For those of us who grew up reading such
mindboggling material it is sometimes difficult to remember that we
live not only in an age when a man walked on the moon – more
than a generation ago – but also in one in which a permanently
inhabited city is being built in Earth's orbit by fourteen of the world's
leading nations. At least 2,500 artificial satellites now orbit our
planet, many of them designed to fulfil a defence/military role; we
are, then, living in an age when we can seriously contemplate the
construction of a 'ring of steel' around our world to repel dangerous

[1] Nostredame, M (trans. Erika Cheetham), *The Prophecies of Nostradamus*,
Century Ten, Quatrain 72, Corgi, 1981, p. 468.

113

objects and take charge of our own future. Inevitably, when all the lobbying and petitioning, debate and financial wrangling are complete, the task of authorising, designing and creating such a system will lie in the hands of our world's foremost scientific, military and political minds. These will be the people who will light the blue touchpaper and withdraw. If, someday, their systems save our world from Armageddon they must surely be remembered as the greatest heroes of our future.

They may be unknown to us at present, perhaps even unborn, or possibly leading political figures of today. But, whoever they may be, theirs is certain to be a role more important than any played by a human in the history of our planet. Quite a feat it will be, but by their thoughts, initiative and determination will the future of humanity be decided. Their task will be difficult, not so much because of our uncertainties about many perceived facts – data is emerging every day about composition, origin and so on. Their task will be difficult because of the technical problems that must be overcome, for in many ways new technologies and innovative uses for existing ones will be needed to hunt and kill a potential Earth impactor.

We know that asteroids are confined within the solar system and generally not too far off the ecliptic plane – so they represent a relatively easy target. Comets are slightly different because a number of them are not bound by the solar system, or if they are they have orbits that last for something like 3,000 years, such as that of Hale-Bopp. The orbit of Halley's Comet lasts approximately 67 years, while Wirtanen, a comet targeted by the European Space Agency's new 'Rosetta mission', has an orbital period of five years.

The problem with identification of a would-be impactor is that it is essential to gather data about the body or core, establish its orbit and be certain that it is going to hit the Earth. If Hale-Bopp had threatened Earth, there would have been no time or technique to save the planet. Hale-Bopp reached us within fifteen months of its identification, and was roughly six times the size of Mount Everest. That would have made a bit of a mess – whether it was soft and crumbly or solid, the energy transfer would have been enormous if it had impacted with our planet.

So there are problems. We have to be able to identify the threatening object early enough, and determine with a high degree of confidence that its orbit is going to bring it into collision with Earth. So that means we really are going to be looking at probably no less than eight or ten years before it hits Earth, with our current state of political and technological readiness, before we will be able to respond. Some of that time will be needed to take measurements over a substantial period of the deathrock's orbit so as to be absolutely sure of its orbital parameters. The gas giants in the solar system – Jupiter, Saturn, Uranus and Neptune – can change these orbital parameters, especially bodies as big as Jupiter. Their gravitational fields effectively pull at smaller bodies, shifting their orbits, in the same way that when a light person shares a bed with a larger person the lighter of the two will roll towards the heavier one.

Let us assume that we have quite a long period of time between initial detection of a threatening object and its eventual impact date: ten years. Assume that it is in the ecliptic plane and that it is going the same way round the sun as Earth.

Once the target has been identified and its orbit charted, the next task would be to develop a method of reaching the target in time to prevent it impacting, and deciding what action should be taken when the vehicle reaches the deathrock. Ray F Turner, chief space engineer at Britain's Rutherford Appleton Space Laboratory (RAL) – an establishment well known for its collaboration with the world's space agencies – has tackled the problem of targeting high-speed objects on many occasions. 'If we wanted to go in the opposite direction around the sun,' he told me, 'a very powerful rocket might give us enough power to generate speeds of 40,000 kilometres per hour. *Might*. But probably less. Then at the best you will have cancelled the Earth's velocity, but little else.'

If the object is outside of the ecliptic plane, at any angle up to 90 degrees, calculations for a launch vehicle to travel alongside the comet and match its orbit would be extremely difficult to arrive at. When Halley's Comet returned to a near-Earth encounter in 1986, Turner was one of the chief mission engineers who designed a probe to intercept it:

> On the Halley's Comet mission, we sent a probe – Giotto – but had the problem that the comet was at an angle of eighteen degrees to the ecliptic plane. Fortunately, the comet was coming in the opposite direction and we were able to launch a rocket at an angle so that we intercepted the comet as it crossed Earth's orbit radius. The snag was that the spin of the spacecraft relative to the comet nucleus was 68 kilometres per second: that is faster than any projectile ever fired by

man. That's hellishly fast – any small piece of debris would cause havoc to the probe's systems. But we had great difficulty simulating the impact speeds, so we couldn't test the defence systems of the spacecraft. That is just one of the likely development problems that you would run into if developing a spacecraft with the aim of deflecting an asteroid.

What we did was to use, in the laboratory, powerful pulses of laser light and an airgun, to simulate particles weighing a tenth of a gram impacting with the probe at 150,000 miles per hour. At that speed such a tiny object will penetrate 7 centimetres of aluminium shielding.[2]

Imagine the problems inherent in these speeds. Not only would a deflection-development team have problems physically simulating the vehicle's journey, in order to test its systems under 'battle conditions', but, actually approaching a target at 150,000 miles per hour, the mission would be distinctly prone to error. You may not achieve pinpoint accuracy, and detonating a nuclear device at the wrong moment or in the wrong position may fragment the object, creating a number of smaller but just as dangerous impactors. Worse, you could trigger a deviation in the bolide's trajectory, possibly shifting its impact site to a highly populated city rather than a remote region. If the object were less than a kilometre in size, such a deviation could result in the impact being far more deadly, creating worse economic and social disaster than might otherwise have been.

To ensure we obviate this problem, the probe would have to move in formation with the would-be impactor before any nuclear device was detonated. Ray Turner's latest project (working with the European Space Agency on behalf of RAL) is the Rosetta mission, designed to achieve just that: match the comet's orbital characteristics exactly, establish an orbit about the comet, and then place a robotic probe on to the comet's (Wirtanen's) surface.

The solar-powered Rosetta craft will map the comet's surface using a visual and infrared spectral and thermal mapper (VIMS), among other instruments, and drop a microprobe on to the comet's surface to analyse its composition. The lander, called Champollion, will use small thrusters to prevent a collision, then spikes will be fired into the icy surface, thereby securing the robot on its piggyback ride through space. That done, a camera will begin relaying images while

[2] Turner, R, Space Science Dept, Rutherford Appleton Laboratory, Didcot, Oxfordshire, interview, 1998.

riding the back of this most fascinating of all celestial wanderers. A drill will probe beneath the comet's surface while an automated laboratory will analyse the chemical composition of the great ball of ice and stone. The lander – named after Jean François Champollion (1790–1882), a 'founding father' of Egyptology and known for his use of the Rosetta Stone to translate Egyptian hieroglyphs (1822–24) – will operate for up to 84 hours on the surface of the comet, a little outpost from Earth, relaying pictures home as it flies through space on one of the most alien and fascinating environments imaginable.

Finally, in October 2013 – eleven years after the mission began – Rosetta will deactivate as Wirtanen makes a scheduled close approach to the sun. The knowledge gained from Rosetta, as well as the proving of expertise and technology, will enable future designers to prepare efficient and effective deflection spacecraft capable of saving Earth from Armageddon.

In that same interview, Turner explained the Rosetta mission design, putting the problem of developing deflection technologies into perspective, as the two technological challenges needed to deliver Rosetta and to deliver a deflection device to a comet/asteroid/ meteoroid are much the same:

> Wirtanen's orbital path is only a few degrees above the ecliptic and travels in the same direction around the sun as Earth. To hit Wirtanen, we have to get out to 5.5 Astronomical Units, so we launch from Earth and get into a slingshot manoeuvre with Mars. We slingshot around Mars; using Mars's gravity to accelerate the craft, come back towards Earth, slingshot around Earth, throwing the Rosetta craft out to the asteroid belt; we come back to do another slingshot around Earth, by which time Rosetta will have fired its engines to allow it to dive under the Earth and change its plane by five degrees, matching Wirtanen's angle about the ecliptic. At that point we will be thrown out to five AUs and we will formate [move in formation] with Wirtanen and will inevitably be able to put a lander down on the comet as well.

If the Rosetta model were extrapolated and employed to suggest operational procedure for a deflector to intercept and 'kill' a potential deathrock it is obvious that it might take eight years to reach this monster. Worse still, the cruise phase is eight years but the planning might last for four or five years, as it did for Rosetta. This, plus detection and verification, and the lengthy political processes needed, would result in nothing less than a fifteen-year timescale. It is possible that we might not develop an exotic spacecraft design for the deflection mission if we are simply launching a nuclear device. If the comet or asteroid was a rogue you might have to tackle it differently.

For instance, in order to send a vehicle over the north pole of the sun, as a scientific mission called Ulysses is planned to do, it would have to do a slingshot over Jupiter in order to do a plane change – Saturn or Jupiter would give you enough energy to do a 'gravity assist' to change direction. A slingshot uses a planet's gravitational pull in addition to carefully timed firings (known as burns) of a craft's rocket engines, to whip round a planet and be thrown off as if a stone from a catapult. Of course, if neither Saturn nor Jupiter was in an appropriate place in space, the slingshot option may not be possible.

Any custom-built deflection system would have to have sufficient in-built redundancy to allow it to maintain its systems for extended periods and to withstand very high or low temperatures depending upon which planets the craft was slung round – Mercury is very hot, Saturn relatively cold because of its greater distance from the sun.

The fact that you might not be able to ascertain the would-be impactor's orbit with any degree of accuracy in time to launch and intercept before the object gets too close to Earth suggests that the 'miss distances' tolerated ought to be significantly increased – perhaps to 500,000 miles or so. This would mean a good deal more deflection activity than would occur if we were to accept a much closer incursion into near-Earth space before launching a defence, because we would deflect more bodies as fewer of them actually penetrate near-Earth space; but the fact remains that safety would be greatly increased, while allowing for a greater margin of error and a second tier of deflective measures.

When the deflection vehicle arrived at the comet, the craft would formate with it and be able to take very specific measurements of the object's composition, perhaps launch a lander or some other form of unmanned probe to do so. It is possible that the nucleus of the object will be a loose agglomerate of rubble, or perhaps it would be composed of iron or ice. Each different form of object would require a different form of explosion, directed in a different way. Agglomerations of rubble would pose a distinct problem: 'What the hell do we do with loose agglomerates?' Ray Turner asked me. It is not an easy question to answer. He continued:

If we were unlucky we might give each of the agglomerates different trajectories, post-nuclear blast, and some of them might be as big as one kilometre in diameter; that could be quite a problem to deal with. It would create numerous targets as opposed to one, if we were to try and have a second attempt to deflect the threat. Something one kilometre across will still devastate the globe or at least devastate a continent.

It is clear, then, that whatever solution is needed will be dependent upon the data gathered by the deflection spacecraft during its approach to formation and the inevitable set-down of a lander or probe on the surface of the deathrock. Any form of deflector will need to be extremely advanced, carrying a number of systems that will allow for reconfiguration of the nuclear blast, type of blast and the specific intended trajectory. In some instances, such as a rubble agglomerate, it may be impossible to achieve the desired effect with one blast and so multiple 'warheads' (a loathsome word but sadly the only term available) will be essential on any such vehicle.

The problems surrounding the use of a nuclear blast to create thrust are possibly insurmountable. Not least of these would be political and ecological resistance to the use of nuclear warheads in space. The international community would be rightly suspicious about the launch of any space vehicle carrying a nuclear warhead. Although nuclear-powered space probes have been launched in the past, they have met with little more than token resistance from environmental pressure groups. The international political community raised little or no resistance to the idea. Warheads are another matter, however. Jasper Wall of the RGO has some sympathy with those concerned about the use of nuclear devices in space:

> One thing that does strike fear into my heart is the concept of building a nuclear-based deflection mechanism. If you really were to build an appropriate mechanism for shooting down asteroids, as seen in the movies *Armageddon* and *Deep Impact*, then you would have the most lethal destruction system that mankind has ever built. I really can understand fears about that idea.[3]

Bearing in mind my discussion of the inherent problems with conspiracies of silence and the restriction of knowledge in the hands

[3] Wall, J, Royal Greenwich Observatory, Cambridge, interview, 1998.

of the privileged few, many countries dependent upon the USA for evidence of a dangerous comet on collision course with Earth might be extremely doubtful that the said comet actually posed a threat. Many might see any US initiative to launch a deflective spacecraft as a ruse, an elaborate cover story, enabling the Americans to place a Damoclean sword over the world, giving arguably the most technologically sophisticated country in the world the benefit of total strategic surprise should it wish to abuse the said deflection technology.

Only by ensuring that the knowledge of some rapidly approaching deathrock on collision course is well disseminated, studied by the international scientific and political community, will we be able to ensure the launch of the 'Earthsaver' deflection vehicle without the risk of war. It is, as I have said, essential that each nation-state have its own near-Earth-object detection system, or better still a world-funded system with nationally based support detectors, in order to achieve some guarantee of impartiality and international cooperation for what would be one of humankind's most audacious and important ventures. I will tackle this problem of how to achieve an international Earthsheild cooperative; first I want to suggest some alternative solutions to the much-lauded nuclear option.

If the use of a nuclear deflection system proves simply untenable in the international political arena, there may be other solutions, though they would absolutely depend upon early detection and speedy interception, as they are long-term solutions. It could be possible to deliver a vast engine capable of delivering different types of thrust to the surface of the deathrock. There lies a problem, however. Each one of the objects on collision course will be rotating. Comets studied to date rotate much as Earth rotates; Halley's Comet, for instance, has a rotational period of approximately 11–12 hours. So, assuming that to be the norm, terminating the spin would present a serious technical challenge. It might be impossible, because it would require a great deal of energy. A thruster placed on the surface without terminating the spin would simply result in the comet persisting along the same, or similar, trajectory but rotating about its axis like a Catherine wheel. If the object is spinning, it might be possible to place a thruster on one of its poles to create thrust in one direction.

A rocket engine would probably be of little use. Chemical rockets are excellent at providing large amounts of thrust over short periods and can provide very large amounts of acceleration. However, a thruster would push a large object such as a 1–10-kilometre comet off course only if it were long-lived and able to deliver thrust over many months.

Achieving such a remarkable milestone – the first step in formating with a comet – will have taken twenty years from concept to landing, with eight of those years being taken up by the journey. That cruise time is potentially the greatest problem we face. The Rosetta mission is scheduled for launch aboard Europe's powerful new launcher, Ariane Five, from the European Space Agency's plush and awe-inspiring exotic space base in French Guiana, South America – a domain where one needs no ability to suspend disbelief in order to imagine that Dan Dare, Space Pilot of the Future,[4] or *Star Trek*'s Captain Kirk, live round the corner. Once Rosetta has launched in AD 2003, it will not rendezvous with the comet Wirtanen until 2011 – a disturbingly long period. A period in which governments might topple, political will might diminish or a world war might be fought. If a deflection technology were developed to utilise such a slingshot (energy-efficient launch from Earth) the probability of the deflector reaching its target would be slim, as might the political will required to continue.

After all, it has been difficult enough for NATO to maintain a steady coalition of consensus against Saddam Hussein since Desert Storm in 1991. Indeed, the coalition has all but collapsed, leaving Britain and America to take action while some of the former coalition members, Russia in particular, have actually condemned the two powers for launching Desert Fox against the Iraqis. Could we expect the political status of the decision makers at the period of launch to remain stable for an eight- to ten-year cruise period (let alone a development period)? I think it unlikely. A case in point is the Strategic Defense Initiative, which has survived the passing of several governments, but has suffered constant reassessment and its structure forcibly changed many times during the 1980s and 90s. Whereas the Republican Reagan was a very keen supporter of SDI, he retired from office in 1988 and, by 1992, the Democrat Clinton was elected,

---

[4] Frank Hampson created Colonel Dan Dare for the British comic the *Eagle*. Dare had many fantastical adventures in outer space from 1950 until the mid-1980s, when he retired from active service and the *Eagle* comic was mothballed.

and was soon heard, in 1993, to denounce the SDI as an offence against international arms-reduction efforts.

No, I believe a defensive system will have to operate in a different and much more responsive way, not only because of political intransigence on Earth, but also because impact warning times are unlikely to be as long as a decade – a matter of days or weeks in some instances. Any effective deflector would, I believe, be launched from an orbital platform, and not from Earth, thus greatly reducing the need to expend energy in order to escape Earth's gravity – an energy requirement that demands a rocket to be capable of an acceleration of 11.18 kilometres per second per second. The further away from Earth an orbital platform is, the less energy would be needed, allowing more energy to achieve faster intercept times and/or orbital-plane changes.

Ion drives may offer a better solution than traditional rocket engines. At the moment a small ion-drive engine on a spacecraft – assuming you had all of the required electrical power, perhaps from a small nuclear reactor – would need 1,500 watts to develop thrust, as ion drives are particularly power-hungry. This new form of drive is only just at proving stage, the first ion-powered spacecraft, NASA's Deep Space 1, having been launched in 1998.

Deep Space 1 (DS1) was designed as an experimental vehicle. Its new ion engine can, if allowed to continue thrusting, propel the ship to incredible speeds. DS1 is scheduled to fly by asteroid 1992 KD on 28 July 1999 and will also perform a fly-by of comet Borreally in October 2001, if all goes according to plan. DS1's engine uses an inert gas, xenon, as its fuel source. The engine develops thrust by emitting electrons from a cathode: a radioactive emitter of the kind used to develop a picture in a cathode-ray TV tube. The cathode's electrons are sent through a chamber of magnets. Xenon atoms are placed in front of the electrons, resulting in collision and invariably in the loss of 1 in 54 of a xenon atom's electrons. This change results in the xenon atom developing a positive charge, becoming an ion.

Thrust is now produced as the newly produced ions hit electrically charged metal grids: the positive–negative charge exerts a tremendous force on the ions, driving them out from the engine into space, at speeds of approximately 100,000 kilometres per hour.[5] Deep Space

---

[5] 'Deep Space 1 Ion Propulsion System Starts Up', NASA PR: 98–215, 1998.

1's engine demands 2,500 watts of power to generate its maximum thrust of 90 million newtons (a newton is the force that would accelerate a mass of one kilogram by one metre per second per second). Potentially, the ion engine will be the most likely form of power used for interplanetary work. Once away from Earth's influence, ion engines can thrust continuously (as long as there is a good fuel supply). They are not good at escaping Earth's gravity but would be ideal to mount on an asteroid in order to steer it away from a collision course with Earth. It is most likely that a powerful rocket engine would be used to launch the NEO-hunter-killer craft from an orbital platform, and an ion engine may well be used to boost the craft to its destination. Another ion engine would then be positioned on the surface of the deathrock.

A very large ion engine might, if fired over a number of years, develop enough thrust to move the hazard away from Earth. But it would need to reach the target long before the target got close to Earth, as the ion engine is more of a marathon runner than a sprinter. Its thrust would therefore generate only a gradual change in the object's trajectory. The earlier that change is initiated the better. If the object appeared quickly, as 1996 JA1 did, being spotted when it was but a few days from Earth, only a nuclear blast could generate enough thrust, with extreme brevity, to push it away.

The nuclear option is not only problematical because of political objections. A nuclear detonation generates a spherical expansion of energy/thrust. If a nuclear warhead were detonated near the surface of an asteroid, a good deal of the energy generated (as much as 50 per cent) might be lost. Only by shaping the blast in some way or perhaps burying the device just beneath the surface of the NEO could thrust be generated efficiently and in one specific direction.

Some notion of the potential of directed nuclear blasts to generate thrust was gleaned at Los Alomos in the 1940s and even then was proposed as a way of achieving interplanetary flight: nuclear-pulsed rockets.[6] This concept depends upon the detonation of a nuclear bomb dosed with an inert gas, such as xenon, behind a spacecraft. The spacecraft would require a collection mechanism, probably a hemispherical shell, to capture and focus the resultant superheated

[6] Martin, AR, and Bond, A, 'Nuclear Propulsion: a Historical Review of an Advanced Propulsion Concept', pp. 283–310, *Journal of the British Interplanetary Society*, Vol. 32, 1980.

gas – known as plasma – generated by the explosion in the field of xenon. The plasma would fill the hemispherical collector like wind in a sail and push the craft away from the expanding explosion. Some form of absorption mechanism, akin to a vast spring, would be needed behind the hemispherical blast scoop in order to ensure a smooth transfer of the blast energy, rather than shocking the spacecraft's hull every time there was a detonation. Once the transfer spring mechanism had finished its recoil and returned to its starting position, another bomb could be dropped and detonated behind the ship, generating another pulse of thrust. Repeating this process would generate enormous amounts of thrust.

Unfortunately, nonproliferation treaties, and the 1967 Outer Space Treaty in particular, effectively prevent the testing of this hypothesis. Nevertheless, the concept of focusing a nuclear blast might be tackled in this way near a dangerous deathrock.

A hunter-killer spacecraft might formate with the NEO and then launch a lander that is capable of installing a large hemispherical focusing shell of the kind envisaged for use on nuclear rockets. The lander robot would secure the focusing shell on to the NEO's surface. That done, a nuclear blast could be released, once again in a xenon field, producing a huge cloud of plasma. Caught by the hemispherical shell, the plasma would generate thrust, pushing the NEO off its trajectory. Some of the shockwave would be attenuated by an anti-shock spring mechanism as suggested for a nuclear-pulsed rocket. Assuming the focusing shell was able to withstand more than a single blast, the procedure could be repeated several times over the period of a few hours or even days or weeks, so as to ensure that the would-be deathrock was pushed out of its collision course with Earth.

Boring into the surface of an NEO may be impossible or too time-consuming, as the asteroid may be composed of solid iron, or alternatively be too soft, being composed of ice and small quantities of rock. I believe this focused form of nuclear-pulse thrust could be the only practical way to employ nuclear warheads in this situation.

Science is badly needed here: research needs to be done on the viability of nuclear-pulsed thrust and the other options when used near the differing forms of asteroidal or cometary objects. My most pressing concern is the threat formed by loose aggregates of mountain-sized rubble. How do we deflect such materials?

Despite fears surrounding the proliferation of nuclear weapons in space, it seems to me that only focused nuclear blasts would generate

124

sufficient thrust, and with sufficient speed, to move an NEO from a collision course with Earth. Whatever the prevailing international political view, this form of device might be the only way of saving Earth at short notice. A spacecraft launched from an orbital platform to target an NEO is a concept in many ways similar to the original proposal for the SDI.

'Building a 150-tonne spacecraft of this sort, in orbit, is only a matter of determination,' Ray Turner told me. 'If the value of saving the Earth is anything, then it could be done. I think you would have to assemble in Earth orbit, as we are doing with the International Space Station at the moment, then send it on its way.'

There may be a way of circumventing the political instability and outcry that might develop following the proposal of such a system. I propose the building of a large orbital system, devoid of nuclear potential, but in every other way fully functional. A series of these devices could be stationed in high Earth orbit in anticipation of the detection of an NEO. When the discussions have ended and the will to take action has been authorised, a specific number of warheads could be launched from Earth, to supply the NEO-killer with its arsenal. The warheads could be fitted in space, either by robots or by astronauts. That done, the device could be launched from its space-based orbital platform, with the backing of international politicians and with a minimum of talk! If an orbital system were primed and ready for action, awaiting only the sting in its tail, the system would become feasible without directly contravening international nonproliferation treaties.

There are signs that this topic may be seen as a special case by world powers, even those ordinarily highly suspicious of Western motives. Most recently, China has argued against the limit on nuclear arms testing. Their Foreign Minister was reported (in numerous media) as saying, 'The door to peaceful nuclear explosions should not be closed, at least not now.' They apparently endorse the belief that a nuclear warhead, delivered by a superior form of rocket similar to the Saturn V, could blast an NEO off course, changing its orbit; though the more cynical among us might prefer to believe that the minister had rather less altruistic reasons for wanting to persist with arms testing.

As we like-minded individuals continue our lobbying process, calling for action on this problem, we must remember that our nation-state governments are faced with a dilemma: either we openly disregard nonproliferation treaties and pursue nuclear-based anti-impact technology, or carry out covert operations. Neither option is attractive. The former would result in international enmity and the latter could have worse consequences: if foreign intelligence services discovered that a nation-state intended to establish a nuclear presence in space (irrespective of motive), economic sanctions or perhaps even war might result. Not a pleasant thought. This topic needs to be handled carefully. Thankfully, the International Space Station experiment, endorsed and financed by fourteen nations, does provide an important precedent in international cooperation of a type never before attempted.

This concept of an NEO-deflection initiative could work if the political and military will existed to proceed. But would the public trust the military to supervise such an operation, or is it more likely that this group would pervert the research into the sphere of 'black (weapons) technology'? I am inclined to trust the military, probably because so many members of my family have served with the armed forces. Perhaps that is naive, but we must all draw conclusions based upon our own experiences. I believe that only the military and its industrial contractors have the energy and organisational and technological skills, as well as sufficiently weighty political lobby, to get this job done.

The European Space Agency (ESA) is not alone in launching an ambitious comet interceptor: NASA already has its own Near Earth Asteroid Rendezvous (NEAR) spacecraft in space heading for the asteroid 433 Eros. The fact that NASA, an organisation with exceptionally strong links to the US military/industrial complex, has launched such a mission is a great precedent for the potential involvement of its defence partners in any future deflection initiative.

NEAR is a tremendous mission launched as part of the NASA director Dan Goldin's Discovery programme, designed to meet his requirements of better, faster and cheaper space missions. NEAR's target, 433 Eros, is a large asteroid measuring 40 by 14 by 14 kilometres which sits in an orbit round Mars, coming no closer to Earth than 14 million kilometres.

NEAR's mission operations director is Mark Holdridge of Johns Hopkins University Applied Physics Laboratory (APL), Baltimore, Maryland. His enthusiasm for the project is reinforced by a clear concept of the difficulties the spacecraft will face and the potentially vital role it may play in helping future designers develop a hunter-

The Near Earth Asteroid Rendezvous (NEAR) spacecraft is on course for a rendezvous with asteroid 433 Eros, its mission target. After reaching Eros, the NEAR will start its orbit about 600 miles above the asteroid's surface, descending to 10 miles during its year-long study.

Impact Earth – a NASA artist's impression of an 800-kilometre asteroid striking the Earth. The atmosphere bursts at the moment of impact, with vast quantities of ejected debris being hurled into space, only to fall back to Earth, filling the atmosphere with superheated particles and dust. Such an impact would be truly devastating, being approx. 80 times the size of the K/T impactor.

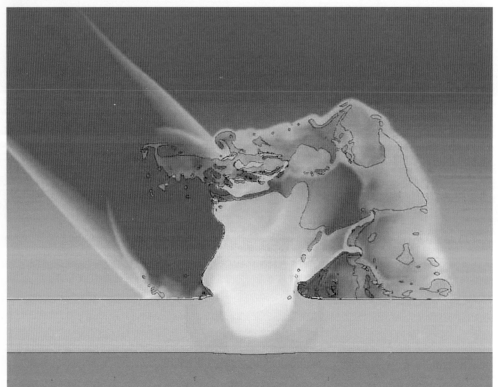

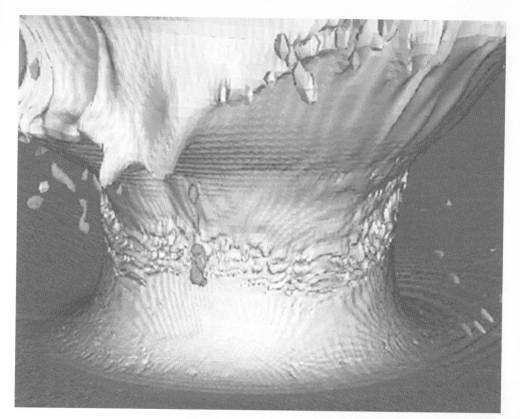

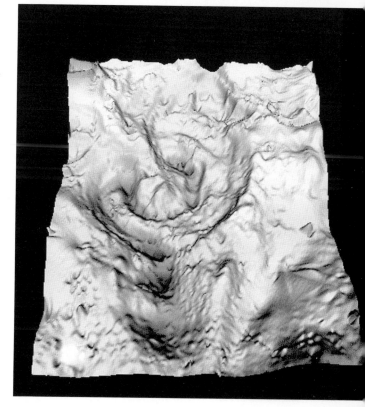

Top left, bottom left and above Sandia scientists, led by David Crawford, performed a computer simulation of a Shoemaker-Levy 9 type comet impacting in Earth's oceans. The 1-kilometre comet weighing 1 billion tons was simulated travelling at the speed of 60 kilometres per second and entering the atmosphere at an angle of 45 degrees. The impact released 300 gigatons of energy and instantly vaporised the comet and 500 cubic kilometres of sea water. A vast tsunami was generated.

Right A computer graphics image of the 65-million-year-old Chicxulub crater, which is approximately 200 kilometres in diameter and buried beneath the modern surface of the Yucatan Peninsular. This map was created using gravimetric data and reveals the horseshoe shape most likely created by an oblique angle of impact.

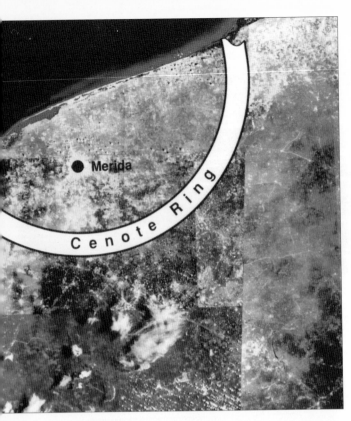

Left This image shows the placement of the ring formation made by hundreds of cenotes – natural wells formed by the collapse of overlying limestone – on the coast of the Yucatan Peninsula. Here the Cenote Ring can be seen imposed on a mosaic of Landsat images, revealing the location of the Chicxulub impact crater.

Below An aerial view of a cenote approx. 150 metres in diameter. The ancient Mayans used the cenotes as sacrificial wells, little realising the true origin of these unusual structures.

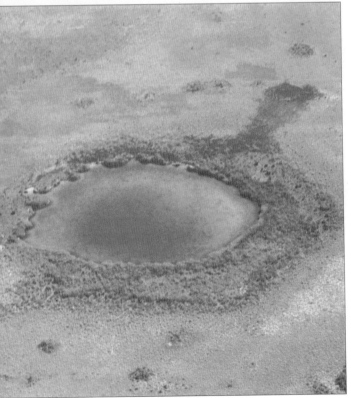

Right These two satellite images reveal the sea temperatures (SST anomalies) on 11/8/1996 (top) and 11/8/1997 (bottom). The 1997 image deviates from the norm, as represented by the 1996 chart, particularly in the vicinity of Indonesia. A huge plume of smoke and aerosol particulates obscures the satellite's temperature readings, ably demonstrating the effects of fires on the atmosphere. The Indonesian rainforest fires, in 1997, are a good model of the likely horrors of a post-impact fire. Cooling clearly took place, over a large area, as a result of reduced levels of solar radiation, thanks to the persistence of the smoke cloud. Even a relatively small impactor (300 metres) would generate similar fires, with analogous consequences – though probably on a much larger scale.

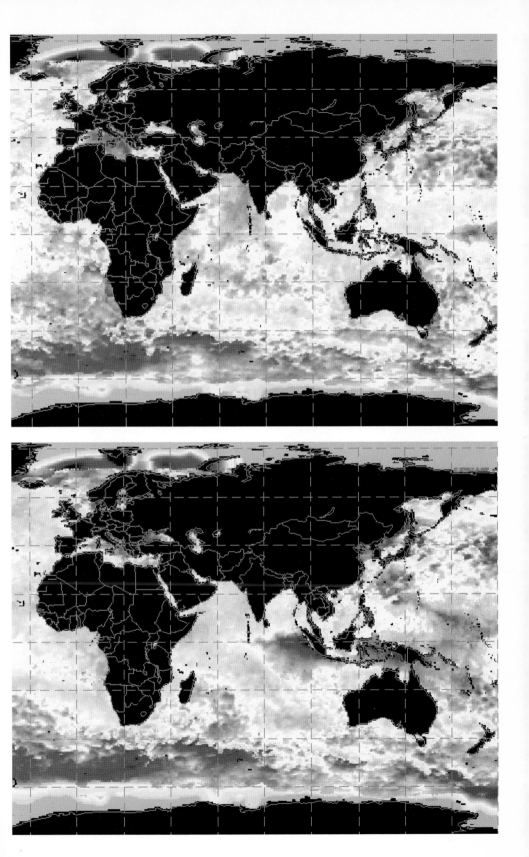

Right A stone tablet from c.87BC is one of the oldest human records of a return to Earth by the body that is now known as Halley's Comet.

Below A comet appeared in the skies in 44BC, apparently confirming the deity of the recently assassinated Roman leader Julius Caesar. Augustus, Caesar's adopted son, minted a commemorative coin bearing a reminder of this celestial omen, and his father's murder, which read 'Defiled Julius'.

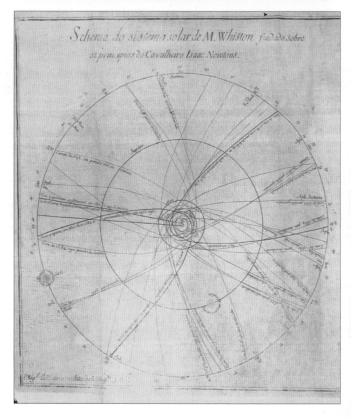

Right This illustration, taken from a book (1757) dedicated to the espousal of rationalism concerning comets and other celestial bodies, which were traditionally seen as omens of doom, depicts our solar system according to the theories of Sir Isaac Newton. The diagram acts as a good indicator of the spread of Newtonian science during the eighteenth century.

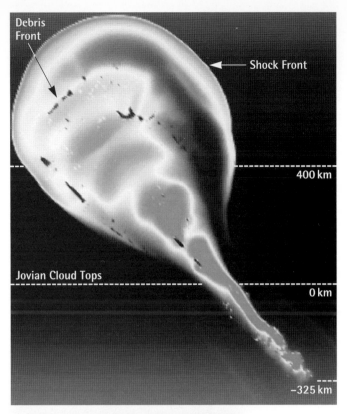

Debris Front

Shock Front →

400 km

Jovian Cloud Tops

0 km

−325 km

Left This predictive simulation of the impact fireball thrown into the Jovian atmosphere by a fragment of Shoemaker-Levy 9 was created by Sandia National Laboratories shortly before the real impact occurred, in 1994. The simulation proved to be stunningly accurate. Depicted as if seen from within Jupiter's atmosphere, 70 seconds after impact, we see the side of the impact fireball as it races away from Jupiter. The simulated fireball reached temperatures in excess of 2,000 degrees Celsius, whilst the impact was calculated to have released 6 million megatons of energy.

Below Gene Shoemaker, co-discoverer of Shoemaker-Levy 9, became the progenitor of the pressure group known as Spaceguard.

Above An image of comet Shoemaker-Levy 9, taken on 30 March 1993, on its final fateful journey towards Jupiter. The comet was pulled apart by Jupiter's vast gravitational field on an earlier pass in 1992, creating a string-of-pearls. The larger central 'nucleus' is actually composed of at least four comet fragments.

Above Comet Shoemaker-Levy 9 impacted with Jupiter in July 1994. The Jovian atmosphere was severely perturbed, displaying impact scars that were larger than the entire surface of the Earth.

Comet Hale-Bopp was first discovered by modern man in July 1995. Seen here over the vast 4,300-year-old stone circle at Avebury, Wiltshire, Hale-Bopp's tangible presence in our skies was, in a sense, an omen, reminding us that great mountains of ice and rock can plummet from the heavens and eradicate thousands of years of human society, learning and history, in a single shattering blow.

killer for an NEO. In essence this first-step mission, he says, is establishing the ground rules for all subsequent missions:

> The rendezvous starts in December 1998 which is when the large orbit corrections are performed to match our speed with Eros. Then the actual insertion burn is on 10 January 1999, which will actually put us in orbit around Eros. We are absolutely confident that we will hit our target. We do perform small daily corrections to keep us on the correct trajectory. Our navigation team is based at NASA's Jet Propulsion Lab [Pasedena, California] and they actually have done some optical and radar sightings of the asteroid. Its orbit is pretty well known; even its spin is well understood. The unknown part is the exact shape and mass of the asteroid.
>
> We have a plan for a successive number of orbit changes over the course of the year. Some of that will be fine-tuned when we first go into orbit when we will be able to determine what its real mass and orbital characteristics are. The nav [navigation] team is going to have to go back and rework our manoeuvres and we'll have to assess how stable the orbits are. It does appear to be a very irregularly shaped object. We will start off with a high orbit and progressively work our way down as we get more daring![7]

Once the NEAR craft is in orbit it will inevitably perform what is amusingly referred to as a 'controlled crash' on the surface of 433 Eros. The craft's landing may well be delayed if mission scientists ask for an extended period for scientific investigation, as the controlled crash may well terminate the mission. It is hoped the craft will land safely on the asteroid's surface, but it is unlikely, as there is little margin for error and the craft was not in truth designed to undertake a powered descent. In essence, it is expected that the mission will terminate when the NEAR craft crashes. If it does not, and its antennae are still able to relay information, perhaps the project ought to be renamed ON! Whatever the outcome, NEAR is an enormous step forward in our efforts to understand the nature of these would-be enemies.

'This is a very focused mission,' Holdridge said. 'The spacecraft design is very focused, which makes it achievable at a moderate cost. That is likely to be a key factor in the creation of any future mission.

[7] Holdridge, M, Johns Hopkins University Advanced Physics Laboratory, interview, 1998.

It has never been done before – orbiting such a small object – so there's going to be a lot learnt both about orbiting a small asteroid and about the nature of the asteroid.'

One of the greatest technical difficulties that the NEAR team and any future hunter-killer mission will face is the minute nature of the orbit corrections that will be necessary: the escape velocity from a small body such as Eros is only around 5 metres per second. Thruster burns could easily throw a spacecraft out of its orbit about the asteroid. The NEAR craft will inevitably achieve orbit at a height of only 15 kilometres from the surface of this tumbling and irregularly shaped mountain in space. This sort of skill will be essential if an attempt is ever made to land a weapon or some other form of thrust-generating deflector on the surface of a dangerous NEO. NEAR is, then, in every sense, vital to the development of any future deflection technology.

Bearing in mind the importance of the NEAR mission and that of the forthcoming Rosetta, it is sobering to consider that plans did not go exactly as expected in late 1998. After a journey of more than a billion kilometres, the craft's scheduled 20 December rocket burn failed only seconds after ignition. The burn was intended to last for twenty minutes, yet this simple failure cost the mission team sixteen months, delaying Eros orbit from January 1999 to May 2000. This simple fact should act as a warning to us all that when dealing with space no amount of planning and preparation can prevent a failure – however temporary – of a major system.

If a deathrock were on a collision course with Earth and our hunter-killer spacecraft failed to reach its target as scheduled, it could result in the death of millions. There would be few margins for error, technical or human, in such a circumstance.

I cannot imagine a more difficult task, or more frightening responsibility, than being mission controller on some NEO hunter-killer mission in the future. The responsibility for countless lives would rest in your hands. One mistake and you might well condemn a whole continent of people to death.

It is clear then that any spacecraft design would need to be incredibly adaptive, enabling the craft to change its orbit to account for irregularities in the object's shape, and to perform a landing or set a lander (or several landers) down on the surface, while all the time relaying information back to Earth. It is highly likely that such a craft would need to be able to make 'decisions' based on the data

128

gathered, as it could take several minutes for signals to be beamed to Earth and a command to be relayed back. This approach was necessary for the recent NASA mission to Mars, which saw a robotic lander explore the Martian landscape. If the lander had been incapable of determining direction and optimum routes it may well have driven over a crevasse while NASA Mission Control's commands were still travelling through space.

Similarly any hunter-killer would need to carry several versions of the thrust generator (nuclear warheads or ion engines or some other form of motive power) as well as carrying enough fuel to be able to re-enter an orbit should any correction result in accidentally achieving escape velocity. If using an ion thruster to alter an NEO's trajectory, the hunter-killer would probably need to operate for an extremely long period (several years) in order to relay accurate information about the engine's function, the nature of the asteroid and its true trajectory – in essence managing the mission for as long as was deemed necessary. In addition, extensive shielding would be necessary, especially if the NEO in question was an outgassing comet, in order to protect the craft from debris. In short, the deflector spacecraft would need to be extremely hardy and also foolproof, as its failure would lead to catastrophe.

In October 1998 Bob Casanova, director of NASA's Institute for Advanced Concepts (NIAC), announced the first round of grant-funding awards for recent projects intended for the long term. One of those projects was simply titled Shield: A Comprehensive Earth Protection System. Dr Robert E Gold of Johns Hopkins APL, the laboratory that controls the NEAR mission, has determined to create a concept for the first truly effective NEO hunt-and-kill system. The so-called Shield research promises to:

> ... develop the architecture, system-level requirements, and the basic specifications for a comprehensive Earth-protection system that will be practical in ten to thirty years and will protect the Earth from objects measuring tens of meters in diameter to a few kilometers in diameter.[8]

A fantastic promise and one very close to my own notion of the ideal orbital solution to the NEO problem. Gold's proposal reminds us that only a space-based detection system would have the necessary

[8] Gold, RE, SHIELD: a technical proposal to the NIAC, 1998.

sensitivity and coverage of space to warn of an impending long-period comet impact while still giving enough time to react and trigger the launch of a deflection device.

Gold's scheme is a real breakthrough in the pursuit of a comprehensive, largely automated, space-based system, free from the problems of atmospheric conditions preventing observation, while reducing the likelihood of human error.

Our study will attempt to develop a total system for detecting and deflecting NEOs. There's a need for an overall system study – how to put together the best ideas of how you might find and deflect them. Not just asking if you should throw nuclear weapons at it or do you throw a rocket against it, but also how do you stage your system? How far in advance do you have to find them? How many spacecraft or Earth-based observatories do you need? Should you combine Earth and space observations?

The main point that we were putting across in this proposal is that, if you find them early enough, propagate their orbit decades in advance, the amount of deflection you need is approximately a change of velocity of seven centimetres per second divided by the number of years ahead in which you have identified it. If you can find out ten years in advance that this thing is going to impact the Earth, you only need seven millimetres per second velocity variation.[9]

Seven millimetres per second is not a lot. That basic premise means that my preference for employing non-nuclear means to ensure deflection becomes more possible, as a deviation of seven millimetres per second could be developed by any number of propulsive systems, including an ion thruster. The size of engine and thrust potential is strictly analogous to the sorts of spacecraft that we have sent out into space in recent years. In short, the technology to deflect a deathrock exists today, as long as detection and orbital calculations are made well in advance. The problem remains that any long-period comet that appears suddenly – with little or no warning – would demand enormous energies in order to deflect it if it were only weeks away from Earth.

Gold's ideas for Shield include the establishment of 'soldier' deflection spacecraft positioned in interplanetary space, at strategic points

[9] Gold, RE, Johns Hopkins University Applied Physics Laboratory, interview, 1998.

in the solar system, thus enabling the rapid interception and destruction of an object at any point between Earth and the outer planets. Gold's aim is to outline a spread of these killer-sentinels, enabling deflection times to be within a year.

The study should reveal how many such robotic soldiers would be needed to achieve that target, as well as suggest their optimum stations in the solar system. There would be two robotic webs. One web of sentinel craft would continuously observe space, each with a 1-AU viewing range over a 60-degree angle, building an incredibly detailed catalogue of the orbits of asteroids, comets and meteoroids. The second, outer web would be composed of the soldiers, which would respond to the scans of the observer robots and, once mission control on Earth had confirmed the observer robot's data, launch a 'killing strike' against any threatening NEO.

Although the Shield study is not yet complete, the ongoing research is something of a revelation. It suggests using a spacecraft such as the NEAR one, which weighs about a thousand kilograms. That vehicle has enough fuel to theoretically deflect a 300-metre asteroid, assuming it were in place on the asteroid a decade before it was due to impact with Earth. It is evident that the 7-millimetre-per-second velocity change is highly achievable with traditional chemical rockets – surely a viable and well-trusted technology – making the whole concept of deflection a good deal more feasible in the near future.

It is obvious that a chemical-rocket solution to the most common impact threat – 30–300-metre-sized objects – would be a good deal more acceptable to the international political community and the environmental movement than a nuclear one. That fact alone should enable organisations such as the ESA and NASA to contemplate the creation of Gold's Shield network. That done, the far less frequent but extremely dangerous 300–1,000-metre-plus objects could be dealt with on an individual basis – inevitably exploiting the nuclear option. This research will, I feel sure, throw light on a gloomy and little-explored problem. It is a heartening thought that NASA's long-term perspective includes serious research into the amelioration of the impact threat in a most pragmatic and achievable way: making its eventual deployment far more likely, I suspect, in our lifetimes.

The political will still needs to be in place, however, even if such a defence system were based on chemical rockets. Why? Cost. The sheer expense of such a project would prohibit its being the child of any one nation, even one with as many resources as the USA. No one could even begin to estimate the true cost of such a venture at this time, but it would clearly be as expensive as, if not more so than, the International Space Station.

Developing the political platform to control such a system – a stable and responsive body capable of taking the initiative to launch a killing strike against a dangerous NEO – would be essential. No one government could supervise a Shield: it would lack the necessary knowledge and expertise; its intransigence would be too problematic and disruptive to the decision-making process.

I propose the establishment of a Rapid Reaction Committee (RRC) composed of representatives of major countries. Inevitably, those that have invested in the scheme would have a major say in its deployment. The RRC would be able – politically enfranchised by its respective governments – to make speedy decisions for action, thereby removing what is normally an elaborate and drawn-out decision-making process. Eliminating the trawl through various layers of government will further reduce the need to deploy nuclear deflection weapons, as the response time will be so much quicker, enabling a killer spacecraft to achieve Gold's 7-centimetre deviation over a longer period, using a rocket engine.

The creation of an RRC and the all-important multilayered detection and defensive webs would necessitate the negotiation of a new 'International Earthshield Treaty', which would obviously supersede the 1963 Nuclear Test-Ban Treaty and 1967 Outer Space Treaty (more formally known as Principles Governing the Activities of States in the Exploration and Use of Outer Space, including the Moon and Other Celestial Bodies), as well as build in exceptions to the numerous nuclear nonproliferation treaties. In essence, I believe that this particular political discussion should begin at once; the sooner the dialogue begins the sooner it will be accepted that this is a real issue to address.

Until such schemes are debated in political and scientific circles, the minimum that humankind must do in order to have a chance of preserving its future is to vastly increase the observation time – scale up the hunt for potentially hazardous objects.

Astronomy needs, fundamentally, to be increased and its emphasis shifted to focus upon the solar system in order to detect these dangerous objects. Unfortunately, astronomers have tasks and aspirations that involve them in the study of phenomena beyond our own solar system, as Jasper Wall pointed out:

Astronomers are too busy faffing around on their own projects, wondering about the state of the universe. Why on Earth should they look for things that are threatening us? We clearly need lots of new wide-field cameras. Not the sort of telescopes being used by astronomers to study distant stellar objects – they are narrow-field. What we need is big telescopes with large collecting areas, but also with wide fields, like Schmidt telescopes. There are a few in existence, but they are pretty small.

We need to equip these wide-field telescopes with large, silicon charged-couple-device detectors. Those are quite expensive but coming down in price. Technologically, using ground-based telescopes to detect NEOs is not a problem. All we need is a network of these big Schmidt cameras, and the staff to run them: that'll cost a few million for each establishment. That's peanuts. These changes would be barely noticeable in the defence budgets of the Western world.[10]

The realisation of this string of wide-field telescopes is the most easily achieved of all proposals. It is estimated that such a network of six ground-based telescopes, linked to a data centre to manage the catalogue, could be established for a total cost of $50–100 million.[11] Compare the cost of such an enterprise with that of a Hollywood movie – now frequently budgeted in the $100-million-plus range. Surely such an important detection system is worth such a handful of 'peanuts', as Jasper Wall put it.

We can all be Dan Dare, the pilots of our own future, if we take positive action to defend ourselves from the impact threat. Not by building futile bunkers in the back garden. No, we can shape the future and ensure that Earth remains a nurturing environment instead of a hostile and deadly one, but only by working together and by voicing our concerns. Such an approach has certainly worked before, on other issues such as apartheid and environmental pollution. Why not for one of the most fundamentally important? I believe that the responsibility lies with each of us, and only by picking up a pen and writing to our parliaments, building a site on the Internet or even expressing our views on the streets can we be sure that we have some measure of safety from what Nostradamus might call a 'King of Terror' from the heavens.

[10] Wall, J, Royal Greenwich Observatory, Cambridge, interview, 1998.
[11] Shoemaker, E, et al, NASA NEO Detection Workshop, 1991.

SEVEN

COUNTDOWN TO COLLISION

Swiftly from the Dark comes Death;
The Stone's pounding sets the cradle tolling as a bell
The gods decree that breath shall be no more;
Snuffing out the gift
A sphere once blue and gold, now dark
None but the Gods shall know what went before.[1]

Is there a worse feeling than loss – loss of any kind, but especially of those we love? It is numbing, all-pervading; the black mists of grief are all-consuming. That sickening weakness that robs one's limbs of energy and the mind of all purpose, destroys the backbone of the tallest and strongest, demolishes the rationality of the most composed and certain. Not everyone fears their own death, though few could escape fleeting thoughts about the means of that death: will it be a speedy dispatch or an agonising one? I have been in some dangerous situations, during the course of an investigation or recce, where I have pondered just that question. I have met many remarkable people in my travels who have demonstrated an awe-inspiring inner calm about their own ephemeral nature. Some have developed this sense of stillness through religious enlightenment, though much more fascinating are those few individuals who truly embrace the concept that a finite span of years defines 'us'.

These supermen understand how our innate knowledge of inevitable death creates ambition, desire, a need to love and be remembered and the reproductive impulse. Yet even those supermen among us can find little inner calm about the deaths of those they love. Anyone who has truly loved must acknowledge that the thought of losing a soulmate, a great friend or a child is intolerable.

Imagine, then, the feeling of certain death, precognition of one's own doom, but worse still to have the certainty that those you love

[1] Broadbelt, MD, 'Portent', 1999.

134

and respect will be lost in the same instant. Could it be worse? Yes, you may live long enough to watch them die. So it could be with an impact.

The British government has designed a system of alerts, a warning, whereby every telephone, every communication system will operate simultaneously. Whether the political will to use such a system exists is another matter, however. The British press have been aware for some time that a number of voices have been considered for this role, not least the well-known actress Joanna Lumley. Hers might be the last voice you hear if Armageddon were to come. If the world's governments feel confidence in the rationality of their people, they may one day use such a system to tell us of the imminent arrival of some biochemical weapon (such as Saddam Hussein's anthrax) or perhaps an impending high-speed comet impact (should adequate detection systems ever be set in place). Some countdown, unspoken but counted out on a wristwatch, or audible via some alert broadcast, may define the last moments of existence. Set your mind to the concept: what would it feel like to know, with such certainty, that the bony fingers of Death await you?

50 seconds . . . Heart pounding, a terrible feeling of loss, helplessness, and fear.

40 seconds . . . All thoughts of love, of the happiest moments, of the saddest moments, are to be irrelevant. No thought will stir again, no concept of love. The people lost to you in a personal history – parents, friends, people who live only as memories – will truly die as you die.

30 seconds . . . Panic? There is no point and yet your guileless autonomic brain functions trigger an impulse to the adrenal gland, which pumps adrenaline through your system. A bodily function, brilliant in its evolved simplicity, designed to generate the ability to heighten senses to run and to protect – irrelevant. Yet it happens, and your frustration becomes overwhelming. The pressure in your mind, the burning of your throat and the thumping of a strained cardiovascular system build to an unbearable crescendo.

20 seconds . . . Each moment might seem an eternity in itself as you wait. Breath held as if bracing yourself in some way will make death easier. Will it hurt? Will the last moments be agony? Or will you lose consciousness? You cling to a loved one and they to you. Yet the arms and body that you love, the smell and feel of the form that gives comfort when nothing else will do, can do nothing to save you now.

10 seconds . . . Your heart rate slows in response to your restrained breath; your mind is highly focused, hearing acute, eyesight brilliant. You wait in silence, still, unable to move.

The countdown ends.

Not a very pleasant thought, is it? These feelings are real; I've experienced them, as have others, in sticky situations. Thankfully the countdown has not yet started and I pray to all our gods that it never does. It is possible and we should, I believe, be ready for that possibility. We should at least take active measures to prepare ourselves mentally as our parents and grandparents must have done when, as during World War Two, death rained from the skies on a daily basis. We must not hide, discussing this problem in whispers in cloistered rooms and ivory towers. We must act. Even cheap short-term measures, those favoured by all politicians, might help prevent the countdown commencing.

Modern warfare, fought by remote control and supervised from afar via satellite monitoring systems, prevents 'unacceptable casualties' – a phrase that presupposes that any casualty is acceptable. In many ways this new form of conflict, as typified by the Gulf conflicts of the 1990s – armchair warfare watched live by viewers around the world – has fostered a feeling of false safety. Death can fall from the sky again, whether it is sent by a madman or a quirk of astronomical fate.

Here is, I believe, a growing threat posed by our blissful ignorance in the face of vast knowledge. It is an ignorance of choice, not necessity. Many of us choose to ignore reality, rather than learn (or even attempt to learn) of the real dangers posed. It is time to grow up as a society and accept our responsibility to our own kind and the creatures who share our planet with us.

We are the custodians, whether by Darwinian evolutionary fluke, impact- and climate-driven adaptation or the whim of some divine being. We are the Keepers of the Gardens. But our immaturity may lead Eden to burn, as we waltz to the tune of reckless abandon, all the while pointing to our ignorance of the threat. Our Keepers' passes are about to be revoked. In many ways I wish there were an alternative, a Neanderthal subset of humankind standing in the wings, ready to take centre stage and make the decisions that we seem unwilling to make. But it will not happen: we are on our own.

We have acted on other matters – environmental, social and economic – so let us act on this threat, a threat that could obliterate all our other efforts.

Expensive projects such as Dr Robert Gold's Shield may be the long-term answer, but seeds could be planted today, the first buds from which would offer some protection and could inevitably blossom to enable the Shield to be built and launched. First we must demystify space.

If we make space 'ours', a commodity to be traded on the world markets, then we will begin to understand that it is but another part of our environment, just like the sea or the land we inhabit. It is a resource cache: far from being an empty vacuum, it is a place filled with materials, plasmas and particles, many of which promise untold wealth for those with enough initiative to exploit them. If we understand how interactive the Earth–space interface is, this shore between the two domains, then we may begin to ready ourselves for the dangers it holds. Just as we have understood the threat of flooding and built dykes and developed great barrier technologies, as on the River Thames in London, so we can develop space barriers.

At least one man on Earth has the vision and drive to make such a historic leap. His name is Jim Benson.

Benson is a fiftysomething industrialist with origins in America's agricultural heartland: Kansas. His remarkable career includes advising President Jimmy Carter on his presidential campaign and the creation of Compusearch, the company that originally developed the word-based search algorithm that eventually evolved to form the search engines used to sift through vast caches of words in computerised databases and the Internet today.[2] Benson has long enjoyed a vision of humanity's outreach into space, believing as I do that only by commercialising and demystifying space can we begin to understand and tackle the dangers that it presents to us.

In years to come, 1998 may be remembered as a pivotal year, as that was the year that Benson etched his vision in corporate legend. SpaceDev Incorporated is already an enormous concern, having bought wholly owned subsidiaries in a brave and innovative vertical integration: not only can SpaceDev design and build spacecraft but it will be able to launch them too.

[2] Landesman, P, 'Starship Private Enterprise', *New Yorker*, 26 October 1998, p. 182.

SpaceDev is a public company designed to generate profit from the short- to medium-term development of space industries. The first of this young but incredibly energetic company's missions will be the Near-Earth Asteroid Prospector (NEAP). As Benson explained to me, this mission could be the world's first nongovernment mission to leave Earth's orbit, visit and land upon another planetary body – in NEAP's case asteroid 4660 Nereus.

> We intend to fly the world's first commercial deep-space science mission, to be launched in April 2001. We are selling rides for scientific instruments on the mission to any group of scientists, universities or national space agencies around the world. No natural-resource company and equipment manufacturer has given any thought to processing in space, therefore there's nothing practical on the horizon. But there may be as many as 100,000 of these near-Earth objects, most of them containing highly concentrated natural resources. There's an abundance of natural resources to be exploited.
>
> I want to make money [by] making history. There's a personal challenge to see how rapidly I can grow a company that helps get humanity off the planet. My main motivation is that, after forty years of watching space and waiting for major breakthroughs that haven't happened, I'm just impatient. I'm in a position in my life, at this point, where I might be able to make a contribution to getting humanity into space. It can only be done commercially, one small step at a time. History has shown us in the last three decades nothing has happened in space because the taxpayer's money isn't there to make anything happen. The only thing that's going to make the infinite space frontier open is commercial success. It's like breaking the four-minute mile: it is mainly a psychological barrier that people don't understand that space is a place and not a government programme.
>
> If we're successful in commercialising space, that might break down the psychological barrier and encourage others to try, because there are limitless commercial opportunities related to space.[3]

Benson's NEAP craft will launch in 2000 or 2001. Make no mistake: he is no crackpot inventor, deluded and blinded by his own zealous views. Benson is a hardened businessman with a first-class brain. He has gathered a team about him who are the cream of the world's space elite. SpaceDev's five subsidiaries include Space Innovations Limited, a UK-based firm which, since its inception in

---

[3] Benson, J, SpaceDev, interview, 1998.

1986, has developed a reputation as a first-rate specialist producer of space engineering products such as low-cost satellite technology and ground stations. The team of forty employees will be used to build much of the NEAP's subsystems; their track record extends to more than a dozen current European missions as well as a raft of space missions being designed around the world.

Most interesting of all Benson's subsidiaries is his Hybrid Rocket Division (HRD). The HRD is currently involved in performing design analyses and computer simulations of diverse rocket propulsion systems and designs. The goal is to develop an optimum launch vehicle that offers cost-effective launches employing new technologies where necessary and the best of the tried and tested where appropriate. The HRD reflects Benson's whole approach: use all available technologies to force down costs and increase profit – something that is seldom on the agenda of state-owned organisations. In many ways Benson is the ultimate pragmatist, but a pragmatist with a remarkable vision: to design, build, launch and command space missions. Space-Dev will make history and, if successful, change our views of space. The world simply does not yet recognise that fact.

SpaceDev has the advantage of insuring its spacecraft against risk, and consequently needs to build its vehicles only to a standard that is insurable. Government agencies such as NASA have no such luxury and consequently they must ensure that the spacecraft have very high dependability in order to minimise the risk of failure. Naturally, this degree of dependability is expensive and SpaceDev will inevitably be able to further reduce its costs to its customers. It seems likely, in this period of budget restriction and increased scrutiny, that former NASA and ESA customers will turn to SpaceDev for a more cost-effective solution to their space requirements.

The Near-Earth Asteroid Prospector is expected to rendezvous with the asteroid Nereus in May 2002 after a four-month cruise through space that will include six close passes over Earth's moon – enabling science to be done, and instrument space sold, which is in addition to the primary mission. The 300–700-kilogram craft will perform a number of operations over a period of approximately thirty days around the asteroid, including the launch of a number of probes on to its surface. The probes will analyse the asteroid and transmit the information to the main craft, which will remain in orbit. At the end of the mission the craft will land on the asteroid. If the NEAP succeeds in all of its mission parameters it will have

139

been a remarkable spacecraft as well as a history-making one. I can only wish it and SpaceDev well in their amazing adventure.

The most fundamental value of SpaceDev's chosen activity is that of the infrastructure design, skill development and rapid deployment of a complex asteroid-interception mission – remember that the Rosetta mission will have taken decades to move from concept to rendezvous. If asteroids can be commercially exploited, vital knowledge will be gained, inevitably fostering deflection technologies. 'Unless we're out there working on near-Earth objects in a commercial and industrial manner,' says Benson, 'we're not going to know how to handle them. It is only going to be literally through hands-on experience that we will be able to control and harness them. As long as we are just stuck here on Earth we're sitting targets.'

SpaceDev could prove to be an ideal model for all future space-frontier companies: a dynamic and aggressive approach combined with a pragmatic rationale that depends on the very concept that space can pay. Certainly SpaceDev is off to a flying start, having built a team of ex-NASA staff and world-class university researchers, in order to fulfil Jim Benson's aspirations.

European nations are keen to leap aboard this new commercial approach to space exploration and exploitation; success by SpaceDev may generate the impetus needed to kick-start a new industrial sector. Although exploitation is a dirty word to many, it is a fundamental factor in any economic equation and I am adamant in my belief that space will become truly commonplace only if the engines of industry are brought to bear in this frontier. Despite a popular belief to the contrary, Europe is very active in this field. I believe that it would take little for existing European space industries to adopt Benson's approach. Britain's first mission to the moon, for example, is anticipated for a 2001 launch. Surrey Satellite Technologies Limited (SSTL), a firm based in Guildford, has designed a microsatellite weighing just 100 kilograms with the aim of carrying out extensive atmospheric studies around the moon. Partially funded by the European Space Agency, the mission is designed to encourage scientific dissemination and to help develop the skills necessary for interplanetary missions in Europe – particularly among its youngest and brightest students.

SSTL's Lunar-Sat will be constructed in one of the most innovative ways possible: using existing technology including silicon microchips of the kind developed for car-engine management systems. The satellite will follow SSTL's other innovative microsatellite successes. Being less than the size of a domestic washing machine, it can easily be launched alongside other, larger, satellites. Using this 'piggy-back' approach, where several satellites are launched simultaneously aboard a single Ariane Five rocket, greatly reduces costs. SSTL's ground station at Guildford will receive and process the data from Lunar-Sat. Already at the technical review stage, it is likely at the time of writing that the mission will enter the construction and launch phase in mid-1999. The SSTL team anticipate that this project will lead to further missions designed specifically to exploit the moon's resources in the next decade.

Developing the skills to design and launch truly low-cost and efficient spacecraft is the key to industrialising space and to the inevitable cost-effective execution of an NEO-killing system. With the development of projects such as Benson's NEAP and SSTL's Lunar-Sat, I feel confident that we will see a major revolution in the push for space in the first two decades of the third millennium.

In September 1859 an innovative American named Edwin Drake, a railway conductor by trade, discovered a vast cache of oil in a Pennsylvania creek. He decided to drill to the depth of 22 metres[4] and his tiny drilling rig, by modern standards, initiated a rush for 'Black Gold'. Forty-two years later on 9 January 1901, the oil age truly began. Using Drake's technology, riggers hit a gusher at Spindletop, Texas. The world's greatest supply of oil (at that time) had been discovered. The most remarkable transformation took place, which saw a mineral slime formerly found only in surface pools, or a substance extracted from sperm whales at the rate of 760 litres per tonne of blubber, suddenly become the economic driving force of the twentieth century.

Just as the oil rush and consequential economic and industrial evolution were unimaginable in the closing years of the nineteenth century, so too is the true enormity of the next evolution in economic activity – space. If we as a society make this step soon, as all signs indicate we will, then I believe that we have a very real chance of defeating this impact threat. Benson may well become a George Stephenson (1781–1848) or Alexander Graham Bell (1847–1922) of the twenty-first century. More importantly, the perception shift that

[4] Atkinson, A, *The World in Our Hands*, Reader's Digest/Toucan Books, 1999.

his success would bring to the people and industries of Earth could literally save the planet. An exaggeration? Ask me that again in 2199, when the Shield has saved Earth from half a dozen would-be impactors, and I will gladly give you an answer. NEAP is expected to launch shortly after Britain's Lunar-Sat. We can but wait and hope that this is the beginning of the beginning.

Although political infighting among supposedly responsible space scientists has not helped matters, I believe the perception shift has begun. As Brian Marsden of the Harvard-Smithsonian Center for Astrophysics outlined in his Foreword, the discovery and announcement of an asteroid catalogued as 1997 XF11 sparked a controversy and revealed dangerous divisions within the NEO-detection community.

The key players in this particular drama were Eleanor Helin and Donald Yeomans of the NASA Jet Propulsion Laboratory's Near-Earth Asteroid Tracking (NEAT) facility and Brian Marsden. All of them are incredibly talented scientists, possibly the best in the world in their particular fields. With an orbital arc of 88 days, scientists were able to develop a preliminary orbit. The result was frightening: 1997 XF11 appeared to be on a collision course with Earth. Target date for impact: 26 October 2028.

Brian Marsden, acting in his capacity of information broker at the Minor Planet Center, duly released this data, calling for more observations so as to verify or modify XF11's apparent orbit. He spoke with the media, informing them that he believed it unlikely that the asteroid would hit – but there was a chance, since the predicted miss distance (on available data) was approximately 42,000 kilometres. That distance is roughly eight times less than the distance of the moon from Earth – perilously close.

Marsden was well reported, his news absorbed and disseminated around the globe. His request for more observations was heard everywhere, especially at the Jet Propulsion Laboratory's Near-Earth Asteroid Tracking facility. Eleanor Helin, an asteroid veteran with 25 years' experience, called for old photographic plates to be examined. Brian Marsden's calculations showed that XF11 had passed Earth in 1990, without being noticed. Thankfully, Helin's team found a plate from 1990 which matched Marsden's predicted passage of XF11. With this new information, the observed orbit of XF11 was increased from its previous 88-day arc. The consequence was a calculation by the Minor Planet Center that redrew XF11's miss distance to 965,580 kilometres, in 2028.

Yeomans was widely quoted as believing that the 1997 XF11 affair should have been handled within the NEO-hunting scientific community, at least until more data was known. He apparently felt that the media, and therefore the general public and governments alike, would be desensitised to the NEO danger as a result of the XF11 scenario being viewed as a case of 'crying wolf'.

I have sympathy with Yeomans's viewpoint, but I cannot agree with it. We laypeople are not children: we can understand error, just as we can deal with threat. When an asteroidal or cometary body threatens Earth, everyone with a voice will be needed to force action and to overcome stigma, dogma and environmental worries. Only by true dissemination and openness – a united front – can this be achieved. Whispered conversations are the most dangerous: they are invariably overheard and misreported. I believe Marsden acted properly – like it or not, the Minor Planet Center is effectively *the* knowledge brokerage, the interface, between the NEO-detection community and the world at large. In order to fulfil its function it should call for information and action in the most productive way possible.

I remain concerned that any future discovery will be suppressed for far too long, owing to fears about mistakes and public reaction. Certainly a catalogue of mistakes would be unacceptable, but warnings should be issued and data called for. Who, for example, would damn meteorologists for warning of a risk of hurricanes, even if there were a good possibility that the predictions were inaccurate? Not I. I am not advocating a publish-and-be-damned attitude, but I do propose a greater degree of openness and frankness.

A demonstration of the value of such a warning can be gleaned from the press reaction to the Minor Planet Center's alert. Reputable newspapers, tabloid and broadsheet both, responded in a responsible manner, as did television and radio news carried by the BBC and other major broadcasters.

Britain's *Daily Telegraph* led its front page with the story, spread over three columns, with a colour plot of XF11's expected orbit and an emotive image of the Earth being menaced by a giant rock. The headline read: 'ASTEROID MAY SPELL DOOM FOR HUMAN CIVILISATION.' The *Telegraph* is a cerebral newspaper, well known for its fair handedness. However, the headline grabbed the readers' attention – it was designed to. So even the broadsheets are not immune to hyperbolae. Nevertheless, the story beneath the headline was much more guarded in its tone, clearly quoting Brian Marsden's caveat that: ' ... the next five years should be spent establishing definitively whether the asteroid is going to hit the Earth.'[5]

The rest of the British press responded in a similar fashion. Even the tabloids adopted this approach, although the *Sun* used phrases such as 'close shave' and 'whizzing past'. There is no reason to suppose that they would hesitate to respond in a similar way again.

America's press responded in much the same way. *Newsweek* offered one of the best warnings yet issued on the subject of impact hazard. Its front page was dedicated to XF11 and read: 'Sure XF11 may miss us, but there are 4,000 like it out there.'[6]

Fears about the XF11 announcement being analogous to crying wolf with the result that little or nothing might be done in the event of a real threat are nonsense. I can understand the concerns, but the news industry simply does not ignore a good story, unless it is specifically told to do so via a government gag – known in the UK as a D-notice. If a news story has 'legs' – in other words is important and will remain on the agenda for some time – it will fall from the front page only to climb in status again as the facts are debated. Once the science was confirmed and debate abated with the conclusion that an NEO was going to hit, the story would achieve pre-eminence until the threat was ameliorated. Facts will out. The media are not some mass of ignorant apes (on the whole at any rate): they are largely populated by people who have an urge to teach, pass on knowledge and disseminate facts. I believe that the quality of reporting of the 1997 XF11 warning was first-rate. There is no personal bias here – I was not involved, as I was on a reconnaissance trip in South America at the time of the announcement.

There is no argument for prolonged silence in times of potential threat. Keeping the world in ignorant bliss is a dangerous game. Remember, the knowledge and skills necessary to detect and confirm the orbit of a dangerous NEO lie in the hands of very few talented people such as Marsden and Yoemans. If information is suppressed for fear of error and reprisal, even among this elite scientific community, vital time can be lost, whereas many minds could make light

[5] *Daily Telegraph*, 13 March 1998.
[6] Rogers, A, and Begley, S, 'Never Mind!', *Newsweek*, 23 March 1998.

work of such a problem. Perhaps more problematic is the public's distrust of the scientific community, so silence where a voice of caution ought to be will only increase this unease between the two groups.

The public fund universities and other science-based establishments with their taxes. If voter distrust and apathy lead to a fall in political support for science activities, the science community has only itself to blame. Give voice to your message in a well-calculated manner and the press should (there is no guarantee, of course) respond in a similar fashion. No newspaper wishes to lose credibility with its readers. Mistakes are made, but they are usually a by-product of tight deadlines and publishing schedules. The dissemination system is far from perfect, but it does work.

After 2028, XF11 may still pose a problem. Many more follow-up observations will be needed, as well as reassessments of orbital parameters following possible gravitational influences of the planets and any future impact with other small bodies. At this moment, it can be said that XF11 poses no danger for at least the next century.

Unfortunately, the release of an alert from Brian Marsden, calling for more information, resulted in a backlash from many of his colleagues. Now, while science depends upon healthy debate, I believe that it is dangerous to argue in public, particularly when the subject matter is a relatively new science and one that potentially involves everyone on the planet. Dissension within a community can do more harm to a cause than any other act. If scientists cannot agree about a problem (or even how a problem ought to be presented) then the public (and governments) often feel that there is no problem. That is the general reaction one encounters, whether one is discussing the impact threat or the BSE crisis. Governments will not take adequate action while there are well-matched but opposing schools of thought. If scientists cannot agree upon the reality of a problem why should we laypeople be concerned? I ask those in the NEO-detection industry to present a more unified front. It is essential if we are to engender sympathy among those with power: the enfranchised mass of the people.

One positive aspect of the XF11 debacle was its effect on the public's perception of this problem. People *are* concerned, much more so than pre-XF11. The increased awareness of the issue is only a beginning, but it does reflect a shift, even if it is diluted by an ignorant response from lacklustre nation-state governments. The USA stands apart as a country that appreciates the potential dangers and value of space.

Warnings such as that issued by Marsden do heighten awareness, but such events are few and far between. A strategy must emerge that will allow the awareness of the impact threat to be constantly in the public mind, eventually crawling on to the political agenda as general green issues did in the late 1980s in Europe. Of course, environmentalism first developed thanks largely to Rachel Carson's (1907–64) seminal work, *Silent Spring*, first published in 1962. Her book raised the first alarm about pesticides (especially DDT), its title referring to an unspecified year in the near future when America's spring may dawn without the voice of its birds. Carson feared that America's flora and fauna were suffering from indiscriminate pesticide use. Her voice was heard most clearly. The people embraced her call to arms and slowly there emerged the environmental movement that we know today, which would be unrecognisable to an earlier generation of conservationists such as John Muir (1838–1914). But it was not born of a single event: Carson's call was echoed by thousands of others around the world. Today, the impact threat must be approached in a similar way, using all devices and techniques now available.

The Internet is possibly the most powerful communication tool on the planet – even more so than television. Information is being absorbed in a new way. The change is happening so speedily that it is easy to miss. Children who were not born when that arch advocate of the Information Superhighway, President Clinton, was elected in 1992 have never known a world without the Net. Yet they use this tool in schools to play, learn and communicate. Reading and viewing habits in the home have changed to accommodate this tool; traditional letter writing is almost dead as far as the youngest of our community are concerned, and they have never known communications without email. Internet service provision is one of the fastest-growing industries in the world. By 2010, children born in 1992 will be enfranchised and able to vote for issues that they understand and are passionate about. The impact threat must be one of those issues if they – and all of us – are to have a future.

I expect that much of this campaign's lobbying and call to arms will be done via the Net – indeed much of it already has been. In many ways the impact hazard is one of the first issues to be born and disseminated largely via the World Wide Web and email. We are at the beginning of a new age of awareness. Visionaries such as Jim Benson point the way towards achieving our goals via direct action. Older, less adaptive people should be the last to be targeted with the impact-threat message. Children, students and young voters should be the primary targets. They are, by definition, more willing to accept new concepts, and are, on the whole, better educated at a fundamental level (despite grumbles to the contrary from disenchanted fortysomethings) than their parents and grandparents could have hoped to be at a similar age.

For example, following the 1997 XF11 debacle, a child being interviewed on the British ITV morning programme *GMTV* astonished his host (Penny Smith) by announcing that he wished to become a mathematician so that he could more properly understand the orbits of asteroids and help stop XF11 from destroying Earth in 2028. This child could have been no more than seven years of age! This remarkable example of a shift in perception within the next generation gives one hope. If we can continue to change perception in the young we may have a future.

There are initiatives operating that have produced noteworthy results; others that point the way for new education and scientific outreach programmes. The National Science Foundation's (NSF) groundbreaking Hands-On Universe Program, based at the Lawrence Hall of Science, University of California, Berkley, is an experiment that encourages children in high schools to gain hands-on experience of astronomy and actually contribute to the science of asteroid detection in the process.

Six asteroid search teams operate at Northfield Mount Hermon School in Northfield, Massachusetts, and in November 1998 three pupils – Miriam Gustafson, Heather McCurdy and George Peterson – discovered a previously unidentified celestial object in the Kuiper Belt. Their astronomy teacher, Hughes Pack, supervised the pupils' scrutiny of computer images taken by the NSF's four-metre Blanco Telescope, based in Chile, and processed by the Berkley National Laboratory Supernova Cosmology Program. Once the three pupils detected the asteroid, the find was passed over to one of their collaborating school teams at Pennsylvania's Oil City Area High School. The Oil City pupils verified the discovery (under the supervision of their teacher, Tim Spuck).

This innovative programme not only introduces children to astronomy but also actually makes them astronomers. These pupils are adding to the sum of knowledge. I suspect that many of those participating in the Hands-On Universe Program will fall in love with the science and some may go on to become career astronomers. The find itself was quite important in its own right, as only 72 objects have been discovered, thus far, in the Kuiper Belt.

'This is a fantastic piece of science, of education, of discovery,' the astrophysicist and founder of the Hands-On Universe Program, Carl Pennypacker of Lawrence Berkeley National Lab and the Lawrence Hall of Science, commented. He added, 'The Northfield students' discovery has shown that all students from a broad range of backgrounds can make solid, exciting and inspiring scientific contributions.'

The NSF is especially keen to encourage this sort of exercise; it is a way to involve the whole community in watching the skies. Clearly this form of initiative could help resolve the understaffing and lack of resources currently directed towards the detection of would-be impactors. The NSF's programme officer, Joseph Stewart, said:

> One of the historically limiting factors in astronomy has been simply not having enough eyes available to inspect all the useful images that astronomers collect, but it's very exciting that these kids are contributing to real science, performing actual science in the classroom![7]

Clearly the Hands-On Universe Program is a stunning initiative that I believe could easily be adopted around the world. It is a programme that I hope national governments, education ministries and even individual schools will embrace.

I find the work by the NSF encouraging. While I am sceptical about the value of government involvement in space projects, other than as an arbiter of defensive technologies, I do believe that other quasi-governmental organisations should follow the NSF's lead and deploy their resources in this most practical of ways. This form of education will enable future generations to see astronomy as a normal and essential practice. Much as our forebears placed a great deal of importance on astronomical observations and astrological predictions, before the age of enlightenment, so we will see astronomy within the solar system next century, albeit for slightly

[7] 'High School Students Discover Kuiper Belt Asteroid', National Science Foundation, PR, 98–79.

different reasons. NEO detection requires a general shift in emphasis within the astronomy community, turning our telescopes on to the solar system instead of deep-space/cosmological observation. It is interesting to note that the images being used by students to detect very faint asteroids in the Kuiper Belt were originated as part of a programme designed to study supernovae. This sort of pragmatic sharing of information, two sets of eyes interpreting the same data for different ends, must be one of the most cost-effective ways to boost the search for would-be deathrocks.

Similarly, little or no NEO-detection work (save amateur efforts) is being done in the southern hemisphere. Observation in this area is vital – Earth is effectively blind to impactors that might hit the southern hemisphere. If a very fast-moving rogue (long-period) comet were to approach Earth in the southern hemisphere, we might have little or no warning. The prevention of such an eventuality is largely in the hands of thousands of diligent amateur astronomers.

The Hands-On initiative is a particularly good example of the sort of project (perhaps also involving adults) that might be employed to detect potentially hazardous objects. The Kuiper Belt is, according to current theory, a likely origin for many of the short-period comets in the solar system. If data such as that shared between the Hands-On scheme and the Supernova Cosmology Program could be disseminated among amateur working groups around the world, we might significantly increase our understanding and detection rate of Kuiper Belt objects and the short-period comets that it emits. Current estimates suggest that there may be as many as 100,000 Kuiper Belt objects that measure in excess of 100 kilometres. Clearly this progenitor of short-period comets is an important key to our understanding of small bodies within the solar system. Any models developed to predict the injection of a potentially hazardous body into the solar system must take the objects thought to lurk in this belt into account. Just as the Search for Extraterrestrial Intelligence (SETI) has instigated a programme of 'out-working', which encourages members of the public to use their computers to analyse data received from SETI to help with their search, so I suggest could potentially hazardous objects be detected using existing data.

When Brian Marsden called for more observational data on 1997 XF11, Eleanor Helin's team was able to use his references to check old photographic plates. If these plates were digitised and made available on CD-ROM or via the Internet, distributed with appropriate software, I see no reason why a SETI-type out-working initiative might not achieve good results. In this way, millions of amateurs could significantly boost the amount of follow-up observation that is being done around the world and would not only make a contribution to science but also raise levels of interactivity with the scientific community and raise awareness of the impact issue, at the same time.

One organisation that has contributed significantly to the problem of raising awareness is the fledgling pressure group known as Spaceguard. Established as a result of the Shoemaker-Levy 9 impact with Jupiter, Spaceguard has since spawned a gamut of nationally based affiliates including Spaceguard Australia and Spaceguard UK. Its primary players are remarkable, and include Jonathan Tate in the UK; their efforts have contributed significantly to the improving status afforded to the problem. Their use of television and newspapers has been particularly effective, and, while their target audience is, in my opinion, too old, they have achieved great things.

Spaceguard has outlined a plan to create a total survey of the heavens, cataloguing every threatening object currently detectable. They suggest that a total of six wide-field telescopes would be sufficient to catalogue all potentially hazardous objects within a 20–30-year period. This scheme would operate on a tight budget: initial capital costs are expected to be no more than $50 million with annual running costs of $10 million. To put these budget requirements in perspective, a typical hotel in Las Vegas, such as the MGM Grand, costs at least $500 million to build; a new hotel planned for the Strip is expected to cost almost $1 billion. James Cameron's Film *Titanic* cost in excess of $200 million to bring to the screen. Spaceguard's budget seems not unreasonable, therefore.

I am sure the people concerned with Spaceguard, being nice chaps, will not mind my stating (in the nicest way possible) that they are perhaps a little too old to whip an army of placard-waving youths to a frenzy! Youth and vehemence of the kind demonstrated by Greenpeace and so-called eco-warriors is needed to bring this issue to the fore: Spaceguard must target and embrace youth if it is to grow and succeed in its aims. I suspect that many within the scientific community and Spaceguard itself will groan at the idea of youthful zealots marching on Parliament. I have sympathy: a comfortable armchair and a cup of hot chocolate are attractive to anyone over the age of 25. Nevertheless, history has shown that it is student groups – the youngest and brightest of any generation – that have secured the reassessment of issues and the acceptance of radical viewpoints by the establishment.

Examples? In 1998 Indonesia was gripped by a people's revolution (I am not asking for a revolution – far from it!) concentrated in Jakarta, the country's capital. Initiated and perpetuated by students, the revolt inevitably ousted President Suharto and kick-started a much needed democratisation of the state. In Britain, the country's road-building agenda was altered radically when numerous groups of dedicated protestors literally took to the trees in a most ingenious way, building small villages in the canopies. To be frank, these protestors were less than popular with local communities, particularly in Newbury, where they tried to prevent deforestation to make way for a road-building initiative, but their persistence led the Labour Party, then in opposition, to embrace a new approach to road policy. Once in office, the Labour government continued with its pledge to curtail road traffic and reduce the number of new roads built in favour of public transport.

The efficacy of student protests in the 1960s and 70s – led by the radical Students for a Democratic Society group – over the war in Vietnam and civil-rights issues are now a matter of history. Protests by the young, myself included, kept the fight against apartheid in South Africa – spearheaded by pressure groups such as Amnesty International – alive in the 1980s. Live Aid, organised by the musician Bob Geldof, pushed the plight of starving Ethiopians on to centre stage – dominating the media and achieving record television and radio audiences worldwide while raising approximately £50 million ($80 million) for famine relief – in 1985. I believe the young will win the impact argument. Only by encouraging existing environmental groups, such as Greenpeace, to embrace the impact threat, and the fostering of a much deeper understanding of the threat among the very young, will this issue achieve pre-eminence as global warming and countless other issues have before it.

There are some useful regular celestial occurrences that can assist in the mission to educate and disseminate. The modern fascination with meteors began in earnest on 12 November 1833, when hundreds of thousands of 'fireballs' were seen across the East Coast of the United States apparently emanating from the constellation of Leo. These Leonid showers occur every year, though they reach storm levels – perhaps as many as 5,000 meteoroids will be visible every

151

hour[8] – approximately once every thirty years, when the orbit of the comet Temple-Tuttle, the engine of the Leonids, brings them close to Earth.

I believe that these storms – a major one is expected in November 1999 – can play a large part in increasing awareness of the Earth–space interface and, although they do not pose a significant threat to environment and life, they do emphasise that currently we can do little but watch debris fall from space. If one can foster understanding of the Leonids, then the concept of much larger space-borne bodies impacting with Earth is a small leap in understanding. The Leonids, along with more than ten other major meteor showers, are discernible, very real, highly visible. One of the key problems about this issue is its separation from our everyday reality: an unrealness. Certainly showers such as the Leonids pose a problem to what little industry there is in space at the moment: approximately 2–3,000 satellites.

Because Leonid meteoroids travel towards Earth at speeds of approximately 71 kilometres per second,[8] they can inflict serious damage to orbital satellites – meteoroids from another meteor stream (known as the Perseids) are thought to have destroyed the ESA's Olympus spacecraft. Large particles can change a spacecraft's attitude due to momentum transfer on impact; similarly, microscopic particles can generate a plasma field on impact which can severely disrupt electronics.

The Roman poet and philosopher Lucretius (c. 95–55 BC) well understood Earth's exposed and vulnerable position when he wrote: 'Even the curass of the sky, which encompasses the world, is not proof against the invasion of tempest and pestilence from without.'[9]

It is a remarkable fact that a concept understood by a Roman 2,000 years before the dawn of the space age still escapes the grasp of many of our world's most influential figures at the beginning of the third millennium.

Despite the evidence and obvious nature of the threat and protestations even from within the academic hierarchy of the military establishment, governments may have little part to play, at least

[8] *The Leonid 1998 Meteor Shower: Information for Spacecraft Operators* (Report), European Space Operations Centre: Mission Analysis Section, 1998.
[9] Lucretius (trans. RE Latham), *The Nature of the Universe*, Penguin Classics, 1952, p. 246.

initially, in the amelioration of this threat. In October 1997, President Clinton cut $30 million from the funding of the *Clementine II* mission and spacecraft. This programme is one of the first missions that could be said to have a specific NEO-killer agenda. *Clementine II* will not kill, but it will pave the way for another generation of craft that will do exactly that.

The spacecraft is designed to test and prove state-of-the-art, lightweight, small-scale satellite technologies, analogous to those perfected by Britain's SSTL. *Clementine II* would be driven by a low-cost 'smaller-faster-cheaper' directive in every sense designed to be constructed and launched on a small budget. The main thrust of the mission, however, is its hunter-killer specifications. The craft is conceived as a means of demonstrating the potential of self-controlling target surveillance and acquisition, tracking, interception and destruction technologies. The project objectives include a desire to launch two or three microsatellites that could impact with near-Earth asteroids, in order to gain greater information about composition and structure.[10]

Probably based on ignorance and deliberate misinformation (by those wishing to secure funding for alternative programmes), the decision to slash much needed funding for what I believe to be a vital research project may slow progress on NEO-hunter-killer technologies to a crawl. If an object were to approach Earth before this restricted research bears technological fruit then President Clinton may have unwittingly signed humankind's death warrant.

This political intransigence simply bears out the point that government-funded space initiatives will not be responsible for taking us 'out there'. Agencies such as the ESA and NASA are vital. They force innovation where unacceptable risk prevents enterprise from doing so. Once precedence has been established and quantum technological leaps made, industry and enterprise can soar. It is highly likely, I believe, that any NEO-killer system custom-built to protect Earth will be made by private consortia in a joint venture with government. Industries wishing to protect their spacecraft, including valuable satellites and perhaps more advanced vehicles designed to exploit space-based mineral deposits, may well contribute to the finance of a Shield-like network. It is inevitable that an international body run by representatives of nation-state governments will control the

[10] Clementine Briefing Paper: Project Objectives, March 1997.

deployment of such a Shield, but it may not be the most likely group to pay for it – at least not entirely.

There are many examples of public–private projects around the world. One of the most famous is the Channel Tunnel – which established a rail link between Britain and mainland Europe for the first time – during a four-year construction period from 1987 to 1991. This rail link was built using government funds and commercial loans, amounting to £16 billion ($25.6 billion [11]), and is operated by an Anglo-French company known today as Eurotunnel. I feel certain that industries with vested interests in space in the new century will understand the value of a Shield-type system, protecting their interests in space as well as those on Earth. They will also look to secure a public–private finance initiative of exactly the kind used to fund the construction of the Channel Tunnel and its rail links.

I am not suggesting that governments ignore the problem. On the contrary, no such system could exist without government funds and military involvement. Nevertheless, in order to secure longevity and stability for a major space project – as the Clementine project reveals – governments cannot be relied upon as the sole source of finance.

As Earth's population increases and covers even more of the globe with cities, it is logical to suppose that we will hear many more stories of impacts and recognise previously misdiagnosed disasters as being of 'celestial' origin. The El Niño ocean current (exacerbated by global warming) has caused floods, drought and bush fires around the globe and is forcing humanity to realise how fragile our world is. Environmental conditions we have accepted as the norm for generations are proving to be ephemeral. Even conservative Britain now accepts that it will face flash floods, Siberian winters and tornadoes such as that seen in the Selsey disaster (January 1998), which caused £10 million worth of damage. Yet the devastation witnessed during these events is relatively minor when compared with the destructive potential of a meteoroidal impact.

We face a tempest, a dangerous beast that stalks us through the night of space. Yet our scientists tell us that the same beast creates new worlds in far-flung systems, 450 light years from Earth: two new

[11] Assuming an exchange rate of $1.60 to the pound.

stars – known as HD85905 and HR10[12] – are being bombarded by comets and asteroids, surrendering their bodies to build planets.

A NASA probe, named after the repentant astronomer, *Galileo*, has shown us that the great rings that encircle Jupiter are born of dust ejected from the planet's four small inner moons, as they are bombarded by interplanetary meteoroids.[13] The Beast is fascinating, all-consuming, beautiful – but, above all, potentially deadly. Humankind must take a step in the dark, embrace the next era in our evolution and in doing so accept responsibility for all life under our care on this world. We must use our capacity to reason and imagine in order to create a wonderful tool that will save our Eden from the tempest of the impact Beast.

Let us hope that our threatened race will not be consumed by the black maw of extinction and that we will prevail where other long-lost playthings of the gods failed.

---

[12] 'Astronomers Find Nearby Stars Constantly Bombarded By Comets', University of California, Berkley, news release, 6 November 1998.
[13] 'Galileo Finds Jupiter's Rings Formed By Dust Blasted Off Small Moons', NASA PR: 98–167.

Part II

ONE

TIME TO DIE

*Location: planet Jupiter, 629 million kilometres from Earth. Date:
16 July 1994.*
Brilliant, blinding searing light. Sound beyond hearing, beyond
imagining. That day the attack began. An attack of unparalleled scale
– not one of malice, or of war, but an attack of purely natural origin.
An attack ordained by Nature herself. Twenty-one destroyers, each
carrying an unimpeachable message to a civilisation that barely
recognised the nature of the messenger, let alone the message itself.

If any human being could have heard the screams of alien air being
torn apart across the gulf of 629 million kilometres, across the empty
vastness of space, that human would have heard the screams of a
planet facing destruction. The great Roman sky god, commander of
lightning, faced twenty-one swords stabbing at his hide. Twenty-one
blades made by his own hand: Jupiter's own power of gravity tore a
great object, of unimaginable age, 4.5 billion years old, apart. This
great comet, known as Shoemaker-Levy 9, became a string of catastrophe, impact after impact scarring the surface of the glowing pagan
god of ancient times. The scars cut deep, prodigious holes, created
huge clouds of debris, on a scale so massive that no member of
humankind could truly comprehend their enormity; for no person
has ever stood in a crater so vast that it could literally contain the
entire planet Earth.

Jupiter's great eye, a boundless ancient storm, staring down on
Earth for all eternity, now had other storms to rival its eternal
presence: great debris clouds, swirling within the atmosphere, the
whole planet a disrupted mass for month after month.

Nature had spoken. Sent a beacon. A vast interplanetary warning.
Only one race could watch and reason; could think on the meaning
of this spectacle. But would that race understand the complexity, the
true import, of this harbinger?

159

Location: Arthur C Clarke Building, Defence Evaluation and Research Agency, Farnborough, Hampshire, UK. Time: 14.00 (GMT). Date: today.

Square and unimaginative, the room was typical of its kind – functional, no thought given to human comfort. It was a briefing room, not some public reception area; a room designed to contain secrets. Officially, research was altruistic and intended to develop new ways of defending the people of the realm. In reality the DERA was an outgrowth of Cold War research establishments, their key function to develop ultimate deterrents, weapons of mass destruction and new battlefield technologies. Only the best worked for the DERA.

The room was lightless; the stink almost unbearable. Cloying, undoubtedly Cuban. Very expensive, just like its owner. He was as dark as the room. Moody, almost murky. His reputation was one of a fierce mind not well disposed to those of a lesser intellect. His emotional state was rarely predictable; his superiors endorsed his presence only because of the unique services he offered them. Few had a brain like his. His disgust of others was tempered only by a need to belong to something other than a library of decaying books. He was human after all, but only just. The interview was as brief as ever. It had taken days to arrange and weeks of patience to finally pull off. He had cancelled three times in the last five days. Now they were both in a locked room.

'This sustained attack on my employers is unacceptable.'

'I thought they would use you to shut me up. I didn't know how else to get to see you,' Wannacheck said.

'I like you less than I like them.'

'Sorry.'

The tip glowed, tracing an arc in the dinge as its smoker shuffled his bulk on the inadequate plastic seat.

'Ask your question.'

Wannacheck knew he had only one shot. 'What is the Bulwark?'

'Electronic mail is a very flexible means of communication,' said Thorbridge.

'Meaning?' Wannacheck sat very still, wondering if his game was up.

'Email is hardly secure. Why would I send anything of consequence via such a medium?'

'You mean, unless you wanted it read by a wider audience than it might otherwise have been?' Wannacheck had always suspected that he was being spoon-fed by the big man. Must have needed a great deal of confidence, on the part of the sender, to expect others to intercept his emails and act upon the data.

'You missed the cricket?' Thorbridge asked.

'Don't follow it. Satellite going down *was* a bit of a bummer, though.'

'What happened, Mr Wannacheck?'

The big head, just about discernible in the gloom, suddenly produced two languid eyes, large and watery. Clearly he had had his eyes closed, as if listening to some hidden symphony, in bliss. He liked to play games.

Was this a test? What about the Bulwark?

'Power failure on one of the solar panels. I wrote that up for the Sunday edition.'

They hit him again – the eyes of a man used to overindulgence, but bright, brilliantly bright. 'I did not take you for a cipher.' Unnerving, almost frightening. Wannacheck had been declared a fool.

The eyes narrowed, then relaxed. Almost sympathetic in their shape. It was an emotional state that Wannacheck had never witnessed in Thorbridge before. 'Perhaps you should take a long holiday. Somewhere where it is always warm in winter.' The eyes were gone. Closed.

The door opened, spilling light into the sepulchre that passed for a room. Of course, the light blinded him. It was meant to. Carefully designed, as always. Even while he was escorted out of the complex, his besuited guides, all four of them – seven feet tall and just as wide – walked so close that he saw little more than his own feet. But he did not need to. He could feel it. He could hear it in the silence all around him. An unmistakable stench that no journalist could mistake.

Fear.

The November winds were strong. He had to cup his hand around the Zippo's flame as he lit his cigarette; he inhaled deeply. The tip, searing hot, burnt through the plastic fabric with ease. The great, grey sheets ran for 500 yards in each direction. Eight feet high, they proved to be a most effective barrier to prying eyes. The plastic flapped against the wire fence, like a great sail, as the wind caught it. No cameras on the outside. Farnborough Airport was military. Still had plenty of civilians living in the area of course. Infringement of civil liberties to have cameras operating outside the complex, or so the local council believed. Whatever the reason, Wannacheck was able to peer through the plastic barrier, or at least the hole his cigarette had made. The hole melted neatly, as he tried to increase its size. The fabric fell away, leaving a ten-centimetre section exposed. Even though the light was fading he could clearly make out the massive bulk of a C5A-Galaxy transport jet, flanked by two 747-400s, each bearing the NATO logo. In the distance, near the terminal building, Wannacheck could see the distinctive bulging head of a Hercules C130-J heavy transporter aircraft, with a large refuelling tube jutting from its nose. The aircraft was configured for long-haul flights. Its navigation lights were active, port and starboard flashing red and green. Although it was in semisilhouette, there was enough light to make out the fact that the aircraft's engines were turning over, even while a boarding ramp was in place. Although the rear of the Hercules was largely obscured by the C5A-Galaxy, he could make out the wheels of tractors, towing large military vehicles towards the cargo-bay ramp at the back of the C130-J.

He reached into the boot of his battered old 735i and opened the gunmetal-grey case within. No one had given the old blue BMW, with optional rust, a second glance. Even if they did, they would probably dismiss him as an aircraft enthusiast. Farnborough was world-renowned for its week-long air shows, which had been taking place there ever since the beginning of the Cold War era in 1948. His old Nikon F4 SLR was still in good order, even after the shocks of numerous violent campaigns, not least following NATO troops on their peacekeeping missions to Somalia. He was unusual, didn't like working with other photographers – felt he had to share the story if his paparazzi colleagues were involved. The bayonet mount on the hefty 600-millimetre auto-focus f4 lens clicked into place. His camera was now a formidable monocular.

The engines were at tick-over, even though they created a tremendous din, not a little painful for the craft's passengers. They boarded. Dubious, almost hesitant, they climbed the steps and entered the cathedral-like cabin of the Hercules. Thorbridge hesitated at the foot of the ramp; he waved his colleagues, thin-faced executives, secretaries and obvious members of the establishment's elite, on to the steps, ahead of him. He looked up at the craft's innovative twisted propellers, whirling like massive scythes, then at the fence around the perimeter of the airfield.

The familiar face stared down the lens at Wannacheck. It was impossible for Thorbridge to distinguish Wannacheck's probing lens from a mass of plastic sheeting, at that distance, but it was spooky enough for Wannacheck to take his finger off the shutter release for a moment. But no more.

Thorbridge looked at the great edifice of Farnborough's military field, took out a huge leather cigar case from inside his raincoat, and then, as if in defiance of all air-safety regulations, lit one of his Cuban stogies. He heaved his bulk up the steps of the boarding ramp and was gone.

The plane arced into the sky, its wheels kissing off the ground with little more reluctance than seemed safe, and banked hard southeast. Wannacheck was sure to take a shot of its registration ident, but somehow he thought it unlikely that the aircraft's flight plan would be filed anywhere that might be accessible to the curious. Plane after plane left the runway, twenty-two over the following thirty minutes; many of them were military, some obviously special charter flights, as they were long-haul jets, all bearing familiar logos.

His car pulled away, its exhaust throwing a stream of vapour into the gloom of the damp winter dusk. His six rolls of film held images of many familiar faces. Some of them were so familiar that their presence proved that one of the most important events in British history had just taken place. He accelerated hard, pushing out along the A327 to reach the M3 motorway, and back to London. He had to talk to Amanda and the Old Man, face to face.

Location: the Tennen Range, North of Werfen, Austrian Alps. Time: 15.16. European Central Date: today.
Jack's leg ached, but he was damned if he'd show it. The ankle again. Pain is a state of mind, his dad had said. True, and right now all his mind could think of was pain. Still, the mountain was there and it had to be climbed. What galled him the most was the thought of being out of the running for the captaincy this year. The team had expected him to take the job – wanted him to. They had been really good about it, but he knew they felt he had failed them. He had played by the rules, but his opponent had not – he hadn't enjoyed being brought down and had lashed out with his boot. No one had seen it happen. Of course. They lost the match and failed to make the county finals. Still, he could do with the extra time. Somehow, economics seemed more interesting than it did before the 'accident'.

The air was damp, a clear note of dew and freshly laundered

forests. The wind, not fierce but still forceful, played tricks on his ears: voices and snippets from some far-distant conversation would break through the unresolved whisper of wind against rock, to form precise words and clear phrases. Perhaps there were other climbers scaling the Alps, or could it be early-evening revellers far below in the nearby town of Tenneck? The great folded range, huge compressed and crumpled faces interwoven with ancient molten material and metamorphic rock, stretched to the horizon and beyond, encompassing him, dwarfing him, reminding Jack of his short-term tenancy on the Earth. He reflected on the enormity of the structure – so digestible in diagram form, yet a barely conceivable mass in reality. While he climbed, the European Alps became his own strange alien world here on Earth, yet offered a very tangible link with the distant Carpathians and Asia's Himalayas. An island chain of stone, apparently constant and yet always changing, moving; subtle faulting and earthquakes in Turkey right across Asia Minor to Iran reflecting the restlessness of the stone behemoth.

He adored Salzburg. Helen had been wonderful, so graceful, sophisticated – totally at home. The second evening of the trip had been little short of magical. They ate atop the rock face that overlooked Salzburg, sitting in a terrace café built in the 900-year-old Hohensalzburg fortress. The light had been fading and, although the weather was cool, seven degrees Celsius, the evening was charged with expectant energy. The Alps, their ultimate destination, seemed to cling to the sky like opalescent azure fabric dropped long ago by some fabulous forgetful god. A female god, it had to be: only a woman could create something so powerful, so beautiful.

Iain pulled on the nylon guy rope, mischievously, knowing full well Jack was trying to take an impromptu breather.

'Yes?' Jack asked.

Iain looked down at them, his face clearly visible, despite the gloom and unmanageable-looking overhang that jutted out above Helen and Jack. His cherubic face matched the red helmet perfectly. Of course, if you had called him a troll he would have been offended, but that was exactly what he looked like. 'Is this where you want to pitch your tent tonight?' he asked.

Helen gave him her most winning of smiles. 'See you at the top, big boy.'

'I do wish you wouldn't flirt with the local wildlife, dear,' Jack

164

said while making an unmistakably rude gesture, with two fingers splayed, in Iain's direction.

'I thought trolls were supposed to be –' Helen began.

'Greedy?'

'No.' Helen tugged on the guy rope suggestively. 'I was going to say well hung.'

Jack grimaced and pulled on the rope, dragging it from her hands. 'After you, dearest. Perhaps the Troll will let you test out that particular urban myth if you get to the top before bedtime.'

Helen smacked his backside as he found a foothold and then, stretching his body out across the rock face, found a handhold. His young, lean body moulded to the rock as though he belonged to a species of mountain goat. His finger ends pressed into spatulate pads, taking his weight as he levered himself up on to the overhang. Although the guy rope, established by Iain, was in place, his life was very literally in his hands. At least he could depend on his fingers. Using only his hands, he hung from the outcrop, a sheer drop of four and a half thousand feet beneath him. He was one with the rock. He hung there a little longer than he needed to, enjoying every last second of pressure, every drop of adrenaline as his endocrine system pumped the natural drug around his body. The winter sun was waning fast, but he knew that in fifteen minutes they would have reached the summit of their ascent, just a training hop, 5,000 feet up at the top of the Austrian Alps, an outcrop of rock close to the Eisriesenwelt – World of the Ice Giants – Austria's 25-mile network of ice caves.

He levered himself up, found another handhold and crabbed sideways. He wished he could have spent more time alone with Helen. He wanted her, wanted to be with her. But she hated all thought of marriage – they were too young, she had said. He knew what she really meant: they were never going to marry – he knew her background. No one who had been through a childhood like hers could believe in the institution of marriage. They could be lovers. That should be enough for any man. But he loved her totally and without question. How was he supposed to deal with that?

They had travelled overland, on the autobahn, hitching towards the Alps. Not just a holiday but a rite-of-passage 'thing', he had called it. They had set off from London, eight days ago. It had taken them four hours to hitch to Dover, all of eighty miles, not a great start. Still, by the end of the first day they had travelled via ferry to

Calais. A trucker carrying a shipment of beer from Newcastle-upon-Tyne had offered to take them as far as Strasbourg – they had reached the German border by the end of the first 24 hours.

By the end of the second day they had navigated 1,500 miles, along the A8, to Salzburg. Their goal was simple – enjoy the delights of Salzburg, then turn their attention to the Alps. They had trained hard for four months, to ensure that they were fighting fit for two weeks of climbing and trekking in some of the most demanding areas of the world. In truth, Jack wanted to be close to Helen, under difficult circumstances. He knew her attitudes to relationships only too well – but he thought they could go further than just 'having fun'. He figured the magic of the Alps, shared hardship, would bring them together on a more permanent basis. She had readily agreed to the trip – she was fitter and more able than he was if truth be told – but she had, to Jack's disappointment, insisted that Iain, a mature student of infamous repute, join them. Jack knew why, but he didn't like it.

'You look good, Jack,' Iain called down to him from his vantage point, wedged in a crevice, his back rigid against the inside face of the crack, his legs acting like brakes.

'Yeh, thanks, chum,' Jack said as he pulled himself over the lip in a straight lift. 'You don't look so bad yourself.'

'Hmm, flatterer. What are you after?' Iain said, affecting a camp voice.

'Nothing you've got to offer,' Jack mumbled as he raced up the face, using his anger to feed his aching muscles.

Location: London. Time: 17.01 (GMT). Date: today.
She had taken his call with more than a little trepidation. He was always charming, especially in the beginning, but it had become obvious of late that he was less interested in her than in the information she could give him. It had not started out like that: even though she suspected that his motives might have always been mercenary, she was sure that he had loved her. Something was wrong, something about the information she had given him. She knew it was important: she had that feeling she always got about strange things, a sense of subdued fear in the pit of her stomach. Still, Peter had asked her to help and had convinced her that she was doing a good thing – after all, weapons were bad, weren't they?

But she still took his call with dread. Peter kept on ringing her at work – it was too obvious, but he did not seem to care any more. She cared. Had to. Her boys were at that age when everything has to be right. If she tried to buy them a pair of running shoes that didn't have the right sort of energy-returning sole, or colour, they would feel rejected and, as often happened, she felt as if she had failed them. More than that, though, they needed their independence from *him*. He, her husband, left her with them. That was a good thing – she loved the boys – but did he have to leave with an older woman? A younger twentysomething, yes – she could have understood that. But the ultimate insult of an older woman! She could have been the boys' grandmother.

No, Amanda needed the job. She would have to present an ultimatum to Peter – either he played by the rules or they were finished. She loved tech support in any case. She had always loved computers. It was one area where she and the boys had a common interest. She was still young, thirty-six, but it had taken her five years to get to the position she was in – against the odds. Computing was still a very male-dominated industry – she had to be twice as good and three times as tough to make it to head of department.

She left Sentinel Online, walked along Arundel Street towards the Embankment and hailed a cab. To hell with the expense. She didn't have time to catch a tube – she had agreed to meet Peter in Soho at 5.15, and it had just gone five.

Location: Rockefeller Center, New York, USA. Time: 12.21 (EST). Date: today.

The rotating doors at the Radio City Musical Hall poured their contents on to the sidewalk. Two of the thirty or so people that emerged seemed to be playing out some drama of their own. As in all cities, no one gave their performance a second glance. The crowd moved out into the plaza that fronts the Rockefeller Center. Many of them were laughing, talking about the two hours of good, bad and plain dreadful entertainment they had enjoyed at the world's largest indoor theatre. One of the crowd dashed out of the building and headed for a walkway that overlooked the outdoor skating rink, which, in freezing November weather, was being enjoyed to the full.

Mark was in the mood for an argument. Jenny, on the other hand, was more inclined to laugh at him than she thought safe to reveal.

'Pet lips don't suit you,' she said.

He turned away from her, scowling. He stared at the grand avenue, its vast glass-and-granite buildings stretching in neat grids to the horizon. Very neat and orderly. Very American.

'I felt like such a berk.'

'Oh come on, lover, why should you be able to do it? I couldn't stand up in front of an audience, never mind tell some gags to a bunch of yanks.' She stroked his arm, gingerly. As he didn't pull away, she reckoned it was OK to worm her hand under his arm, then the other, and cuddle him. 'I thought you were fab.'

He pulled away, shrugging her off in a mock attempt to put air between them. Of course he was only fishing for compliments. He pretended to be immensely interested in the activities of a graceful teenager, who managed a pirouette on the ice for a full twenty seconds before being knocked unceremoniously from her feet by a clumsy middle-aged woman. The impromptu attacker's feet seemed totally at odds with her skates, while her shocking-pink tracksuit seemed only to reinforce the impression that she was anything but familiar with the concept of gracefulness.

Jenny smiled and threw her handbag, full of his essentials, on to her shoulder. 'All right, that's it. We are going to Macey's and we're going to blow serious amounts of cash.'

He looked up at her from a heavily creased brow. He tried to look fierce, but her smile teased and rendered his whole mood hopeless. 'That's what holidays are for, right?'

'Right!'

She grabbed his hand and tugged him. He staggered towards her, smiling at the silly red bobble hat she was wearing, and planted a fast kiss on her cold lips. 'All right, gnome. No more gags and embarrassing contests.'

She laughed and snuffled his frozen Roman nose. 'Right, just lots of goodies.'

He stopped her as she walked towards the pedestrian crossing. 'Kiss,' he said. She kissed him again, this time lingering, more tender. He felt the warmth of her tongue on his lips, and breathed in through his nose. That close he could smell the conditioner on her clothes mixed with the odour of woollen mittens and Amirage. She smelt rather old-fashioned but somehow just right. Anyway, he was an old-fashioned sort of bloke.

'Well, Mrs Kingston,' he said. 'I'm shocked. Married people don't do tongues.'

Jenny smiled as she put her arms round him. She watched his eyes

168

change their shape as she slid her hands down to his buttocks. Her eyes, always vaguely reminiscent of an aristocratic cat's, became positively predatory. 'This Mrs does.'

Location: the North Sea, 40 nautical miles SE of the Shetland Isles. Time: 17.35 (GMT). Date: today.

The *Petrushka* cut through the harsh black seas at twenty knots, a barely discernible hulk against the night sky, throwing white foam bubbling and boiling to either side of its rusting bow. The ship was in British territorial waters, illegally of course, but no one on board cared about that – not unless they were caught. They had their bounty. The others had failed, but they could still intercept them, unless the *Petrushka* outran their ship.

Alice Holding was too young to care about protocol. She was skipper on this trip and was determined to stop the Russian ship escaping before the British coastguard arrived to intercept. Her bridge crew, all far more experienced but perhaps less adept, were scared – she was pushing the ship to the edge of its abilities.

The wind tore through the bridge as the mate took refuge from the gale.

'Force eight, northeasterly. It's getting really wild out there, Skip,' he said, eying the rest of the bridge crew, realising they were uncomfortable with Alice at the helm. Bobbie Peters, the coxswain, was beside himself. He had worked alongside his father on trawlers until he'd entered his mid-twenties, by which time his father had seen his livelihood – his whole way of life – eroded by indiscriminate fishing by large-scale factory ships. Bobbie, now in his late thirties, was a keen volunteer and a born helmsman, but, like the rest of the crew, as far as he was concerned, Alice Holding was an untried captain. He looked at the mate and raised his eyebrows as Alice pushed the throttles of the eight-cylinder Gardener diesel engines to full power.

The voice of the winds roared across the seas, straight from the throat of the Arctic Ocean, across the vastness of the Siberian steppes and out from land over the North German Plain. The wind is forever savage in this territory, as those who live in the Isles of Shetland and Orkney can testify. The black seas lashed at the 1,500 tons of the *Rainbow Warrior III*'s bows. Metal against water might seem like an easy contest, but the oceans are more than a match for any artificial hide. Swirls, mountains of cold energy, bulged into the sky,

169

their broken tops torn into crests by the knife edge of the winds. The *Warrior* cut through them, yawing badly to port and then to starboard.

The crew were silent. The radar's unmistakable voice echoed around the bridge as its radio signal bounced back from the giant hull that could only be the *Petrushka* and her fleet of killers.

'How's she laying?'

Ronald den Hartog, the ship's second mate, glanced at the radar screen and made sure of his findings. The giant hull sat near the centre of the first concentric ring on his radar screen. 'She's five miles on the starboard bow, due northeast.' The Dutchman looked over for confirmation that the skipper had heard him. Nothing. Damned irregular. His only clue was Alice's throttle hand, pushing the levers against their housing – forcing every last drop of power into her crusade. He ran his thick fingers through his brown mop and wiped his brow. The ship lurched badly as the skipper threw the ship five degrees to starboard, straight into the waves. Den Hartog glanced at the mate. The first mate moved over to den Hartog's position and leant over to look at the glowing radar screen.

'What's her speed?'

'Twenty knots, Sam,' den Hartog said, glancing up to look into the mate's eyes. He saw worry there. 'He's pushing her . . .'

Sam Macintosh, a 43-year-old veteran, stared at the back of the skipper's head. She had one hell of a reputation. He only hoped this particular obsession didn't cost them all their lives.

'Can they see us, Ronald?' she asked.

'Not likely in this storm, Skip. Too much ground clutter. We're too small a fish, but she's a mountain on the sea – you'd have to be blind to miss her.'

'Call it in, Sam,' Alice said. 'Don't pull any punches.'

'Calling it in, Skip.' The first mate took his place at the radio console and flicked the OPERATE key. He checked the frequency chart, adjusted the digital tuner to 67 VHF, 156.375 megahertz, and hit the hot key. 'Aberdeen, Aberdeen. Aberdeen Coastguard, this is *Rainbow Warrior Three*. I say again: Aberdeen, Aberdeen. Aberdeen Coast Guard, this is *Rainbow Warrior Three*. Over.' He released the hot-key toggle and listened to the static.

Within seconds he had his answer.

'*Rainbow Warrior Three, Rainbow Warrior Three*, Aberdeen Coast Guard, Aberdeen Coastguard receiving. Over.'

'Aberdeen, Aberdeen, this is *Rainbow Warrior Three*, position –'
Sam glanced down at his navigation console and took a reading from
the Decker Navigator, and then glanced to his right and confirmed the
reading with a calculation from a Global Positioning System (GPS)
decoder '– twenty miles southeast of Foula, twenty miles west of
Fitful Headlands. Latitude sixty degrees, seven minutes south. Longitude two degrees, two minutes east. Over.'

Bobbie recognised the sense of urgency in Sam's voice – he was
worried. He looked at the skipper. She was intent, focused and
apparently totally in control. She was a powerful figure, square
shoulders, a mane of blonde hair, tied back a little too severely with
a tatty old scrunchy. She was at least five foot nine of solid muscle.
But she was beautiful. Simply stunning. Although her angular face
was set, determined, it was not hard – it was generous and open.

He looked back at the mate as the coastguard responded to his
send.

'Twenty miles southeast of Foula, twenty miles west of Fitful Headlands. Aye. Latitude sixty degrees, seven minutes south. Longitude
two degrees, two minutes east. Acknowledged. What is your
message? Over.'

Sam gritted his teeth. 'We are in sight of Russian factory ship
Petrushka. She is in breech of whaling embargo in British territorial
waters. Requesting Fishery Protection. Repeat, we are in sight of
Russian factory ship *Petrushka*. She is in breech of whaling embargo
in British territorial waters. Requesting Fishery Protection. Over.'

'Acknowledged, *Rainbow Warrior Three*. Stand off. I say again:
stand off. Over.'

Alice laughed and glanced over at Sam. 'Screw that. I'll be damned
if I'm going to wait for the big boys. It'll take them hours to get a
frigate out here. She'll be well clear by then. Tell 'em how it is, Sam.'
She stared at him, hard. Bobbie watched the exchange with interest:
Sam held her gaze and she his. Bobbie recognised a stand-off when
he saw one. It should have been Sam's ship, but she was bright. She
was a northern girl. Had made a name for herself in the US, skip-
pering an ecology boat patrolling the Hudson River and sailing on a
whaling research vessel. Besides, HQ wanted a woman at the helm
– good for PR. Of course, she might even have been the best 'man'
for the job.

'Aberdeen, Aberdeen. Have received, have received. Negative. Say
again: negative. *Rainbow Warrior Three* closing now. Repeat:

Rainbow Warrior Three closing now. Over.' He glanced at Bobbie and smiled. At least they would see some action tonight.

The Aberdeen Coastguard operator was nearly frantic. 'Negative, negative, *Rainbow Warrior Three*. Stand off. I repeat: stand off. Over.'

Sam looked up at Alice. She turned and winked at him. 'That should put a rocket under them.' He smiled at her, despite himself. 'Hit the *Petrushka*, Sam. Let them know we're coming. Don't let them forget it.' She glanced over at Ronald. 'Better get the Zodiacs ready, Ron.' She nodded towards the port door. 'We'll need them in the water as soon as we're within visual range.'

The *Petrushka* had once been part of the Soviet Union's great fishing fleet. She and her catchers had delivered a lot of food and oil for the USSR. Serge Pavlovich had served aboard her for 38 years. No skipper had been more respected or distinguished by his achievements. A glorious past of no relevance to today's harsh realities. He and his crew had broken maritime law, again. It was a matter of survival. The black market brought healthy prices in the East for whale products, not least in his own impoverished nation – Russia.

He was long overdue for retirement, but his state pension no longer existed – it was worthless, just like the currency that he used to be paid with. A 72-year-old captain. He hated his life, the state that his once great nation had fallen into. He had endorsed the reforms, believed in them, but the harsh reality had been mismanagement and hungry bellies. Now his daughter had a child of her own and her husband depended on the *Petrushka* as much as he did.

Joseph Konrad, captain of the *Vostok*, thought of his father-in-law, on the bridge of the *Petrushka*. He dreamt of replacing the old man when he finally retired, or died behind the wheel. He had never known any other way of living. He had entered the job market at seventeen and come aboard the great ship eight years ago and was now captain of one of the ship's catchers. He loved the thrill of the chase, firing rocket-powered harpoons into the hides of the great creatures. He did not think of the whale's feelings, its ruptured beauty, only the thrill of the chase and the satisfaction of earning a living – however illegal.

The catcher had snared a whale earlier that day. Inflated its body with air and flagged it for the *Petrushka* to winch aboard, via the factory ship's stern hole. He had even come alongside and boarded *Petrushka* in order to see the great minke whale. He had joined the crew as they flensed the whale's hide – severing its flesh and plying it from its muscle structure using curved flensing knives. He was certain the great creature – so pathetic out of its own environment – was still alive when they slit it open.

Location: Twickenham, London. Time: 19.10 (GMT). Date: today.
The noise droned on and rumbled around in Amanda's aural system. Her thoughts mingled with the sound of the well-modulated, trans-mitted voice, her subconscious mind filtering and filing the words in a most conscientious manner. ' . . . REFUSED TO BACK DOWN IN THEIR STRIKE OVER WORKING CONDITIONS. THE AUSTRA-LIAN GOVERNMENT HAS INSISTED THAT IT IS WORKING ON A SCHEME TO INCREASE RECRUITMENT AND INVEST IN NEW TECHNOLOGY IN AN EFFORT TO EASE THE WORK-LOAD OF THE NATION'S AIR-TRAFFIC CONTROLLERS. THIS IS INDEPENDENT RADIO NEWS FOR . . .'

Amanda lay there, immobile, in a heap of clothing and sullied material that had been her neatly made bed. She could still smell his sweat, his aftershave. She wondered where he was, why it was that this story was so big that he had to leave her when she needed him most. She needed him around, his certainty about his career, his self-confidence. He had become her rock. That was bad. It meant that doing what she had to do would be almost impossible – but Peter had to go. The boys could not handle another man in their lives right now. They were still grieving for their absent father – who might just as well be dead as absconded, as they never heard from him, or had any contact of any kind. It was almost as though he were denying that the last twelve years had ever happened.

She reached for her mug of tea at the bedside. The room was murky, illuminated only by a faint orange glow from the street lights outside the thin curtains. She made a mental note to do something about them, maybe buy those velvet ones that she had seen at Marks & Spencer. If it weren't for the boys' school trip to La Roch-elle, she would have done it already. Still, at least the boys would not be denied the experiences that she wanted them to have. She was quite determined about that. Hoped they had had a good night tonight, out at the local cinema with their friends.

Amanda sat up, pulled the duvet up to her chin in an effort to keep warm – the temperature had dropped. She had noticed, on their way back to her house from Soho, that the sky had been completely free of clouds, a winter anticyclonic evening; could get really cold tonight. She sipped her tea and reflected on Peter Wannacheck. What had he said? A conspiracy of silence?

Wannacheck drove as though he were on autopilot, racing through London's empty night-time streets as if they didn't exist at all. The BMW purred as it travelled along Victoria Embankment, running parallel with the majestic Thames river front. He glanced across to the South Bank, the concrete-and-glass rectilinear structure that housed the National Theatre, and beside that the London Television Centre and the distinctive brick, plaster and glass oddity that formed Oxo Tower Wharf.

'I don't get it, Anna. How can his press secretary be unavailable for comment? That's his job.'

Anna Mitcham's voice was tinny and slightly staccato, thanks to Wannacheck's antiquated hands-free cellular phone system. 'It's damn peculiar, Pete. Everyone on the *Six* was spitting blood. We lost the interview with the Home Secretary with no warning and no apology.'

'A total no-show?'

'Yeh, not like the old smoothie, is it?'

'No, but I wouldn't hold your breath for a personal apology.'

'Why, what've you heard?' Anna knew better than to expect him to answer her directly. But he never rang her unless he felt he was on to something extremely big. Almost as though he were looking for personal and professional approval of his success. Despite the fact that they had been close, they had never been lovers – something she struggled to understand. They were great mates and there was a strong attraction, but nothing had ever blossomed. Maybe Wanna-check's ego had been bruised when she got the deputy editor's job for the BBC's main early-evening networked news bulletin. He had left the BBC, rashly, most people thought, and had gone freelance. Still, she hoped that one day . . .

'It's more what I've seen that matters,' said Wannacheck.

'Are you going to Jackson with this whatever-it-is tonight?'

Wannacheck paused. He knew that she would help him break the story with the BBC, but he wanted to do this on his own. Besides, if he was right, the story wouldn't be broken – at least not by conventional means. 'Yeh. I'll call you later about it.'

'OK. Don't bother trying to contact anyone at Downing Street, though, Pete. The whole place has gone tight-lipped. All the news from Number Ten is weirdly spun – no big successes, no targets or stories, just business as usual.'

'Did you try talking to one of the party rent-a-gobs?'

'All the backbenchers are neatly tucked away with a jar of whisky and a stick up their backsides.'

'You think the party whip's been out?' he asked her.

'And then some. All the lobbyists are screaming because they can't get to meet the MPs and the next two days of the PM's diary have been rescheduled. I'm certain my editor knows something about all this, but he's keeping schtum . . .'

Wannacheck froze as he realised he had just driven through a red light at London Bridge. He stamped on the accelerator. The big car's 3.5-litre engine roared as its automatic transmission went into kick-down, throwing all of its power and torque into the drive shaft. A gunmetal-grey Sovereign Jaguar with diplomatic plates roared passed his tail, at right angles, as he cut across its path, the extra acceleration just taking him clear of the Jaguar's nose. He did not speak again until he had slipped across the junction and had put some distance behind him. He could still hear the blaring of the big cat's horns ringing in his ears.

'I'd better go, Anna.'

'Sounds like you're having fun.'

'Something like that. I'll be at my desk in ten minutes if you want me.'

'Now there's an image to conjure with.'

His car screeched across the convoluted junction that controlled the traffic flow over the ornate, but overused, Tower Bridge, and down into East Smithfield, then the Highway and almost immediately right into Pennington Street, the headquarters of News International, otherwise known as Fortress Wapping. Wannacheck was a regular there: security were used to his frequent comings and goings at odd hours. He had a desk, despite being a freelancer – a sort of special privilege extended to him by Jackson, editor of *The Times*.

The place was buzzing – a full house. There were no spaces in the car park. He avoided the grand new atrium entranceway, with its moving staircase and marble floors. He left this grandiose path to the corporate types, preferring to slip in through the side entrance, via the loading bay. The corridors, normally rather quiet by 19.00 hours, were buzzing with staff – but there was no chatter; the atmosphere was intense. He took the lift to the fifth floor and walked through the open-plan office. Everyone was there, even the Sunday edition staff. He got the same sinking feeling in the pit of his stomach that he had had earlier that day, during his conversation with Thorbridge. A tightening all the way through his system: his stomach was so taut he thought he would throw up. What the hell was happening to him? He felt like a fool walking in there. It was obvious, from the look on the familiar faces, the total absence of usual joviality, that they all knew something that he did not. The main office stank – a mass of stale coffee, hot dusty computers, breath and takeaway food. The staff had obviously been there all day, without a break.

He stopped outside the Old Man's glass-partitioned office. Jackson was a notorious bastard, but Wannacheck liked him. Mostly because he was a notorious bastard too. Sally, Jackson's tiresome ever-ready assistant, was notable by her absence. He glanced at her normally pristine desk: it was awash with paperwork and Post-it notes bearing Jackson's scribble. She must be ill, or maybe she had slipped out to pick up a little something to help Jackson get through the night. He knocked and walked in, feeling as if he was about to tell the Old Man that his favourite mistress had just died. If Wannacheck's gut feeling was right, the Old Man already knew what was going on.

Location: North Sea, 60 miles SE of the Shetland Isles. Time: 19.25 (GMT). Date: today.
Two of the small rubber craft cut across the great iron beast's bow. Alice Holding whipped the wheel to its lock, throwing her Zodiac powerboat hard to starboard. The giant *Petrushka* did not seem perturbed. Holding, her face rock-hard despite the freezing winds, cut so close to the great hull that her outboard ground against the rusting hulk as she spun off, barely avoiding a collision.

She knew she was winning. Whatever the consequences, she had the *Petrushka*'s skipper worried. She could feel it. He might laugh in the face of international fishing agreements – piracy was a hotly debated issue – but would he be so happy to face a manslaughter tribunal? The way she piloted her Zodiac, he had to know that she was prepared to go that far, to risk her life, if it stopped his fleet from sailing again.

Ronald den Hartog, Holding's second mate, turned his Zodiac round in an ellipse and targeted the *Petrushka*'s catcher vessel, the *Vostok*. The Dutchman, well used to handling the vibrant outboard-engined speedboat, sailed skilfully to the *Vostok*'s stern. His boat was thrown around by the 15,000-ton vessel's wake, but he rode it out like some aquatic broncobuster. He stood, absorbing some of the impact shock with his knees as the little boat was thrown into the air, jumping the wake, landing, and then leaping back into the air. The Zodiac's Aldis searchlight, mounted alongside the radio mast on a bar above den Hartog's head, traced an arc in front of the boat, illuminating his path around the vessel's hull. He moved out from the rear of the ship, staying close to the hull at all times, intent to overtake and dog the ship without the skipper realising he had been targeted.

He saw an entry point ahead. Something had hit the water – vapour and water had been thrown skyward in a coronet. Something heavy. Den Hartog nudged the Zodiac ahead, his head buzzing with adrenaline, his nostrils filled with the sickening stench of the *Vostok*'s ancient diesel engines, spewing out soot and engine oil. His arms ached; his knees ached. His body was taut, fully alert, in the genetically programmed fight-or-flight response that human evolution has so far maintained from its bestial ancestry. His hands were raw; his knuckles were locked, and had been for nearly an hour, round the wheel of the little boat. Another watery crown – the crew were dumping something heavy, something very heavy, over the side of the catcher. There, above him, he could see the crew. They were throwing it at him. A huge drum, a fuel drum.

Instinct took over. He gunned the engine, and threw the wheel over, forced to tip the bows into one of the twenty-foot waves instead of riding over it as he had done previously. A great surge of black water poured over his head, half drowning him. The Zodiac yawed wildly, but he held on. He ripped the wheel into the yaw, righting the boat, but he had lost the throttle control as the wave had hit. The revs were low and he had lost torque. He grabbed the throttle and locked it down, ramming the knuckles of his right hand against the bulkhead as he forced the outboard to deliver. But the wave had him. The Zodiac was picked up by a great unforgiving hand, whirled and tossed. Its semirigid hull smashed into the *Vostok*. Ronald den Hartog knew he was fighting for his life.

Konrad gave the order. Obeying their boy-captain, the six hands heaved another fifty-gallon can on to the handrail. Their hands were frozen by the unremitting winds and icy seas as they lashed the deck. Joseph Konrad, enjoying the adventure, revelling in the conflict, screamed at them, his face pouring with sweat, his throat hoarse as he screamed to be heard over the screech of the arctic winds.

The barrel fell.

Den Hartog's hand grasped for the throttle. He was inches away, his left hand still locked to the wheel. He heard the sound of the *Petrushka*'s siren blasting through the seas like an old-style foghorn.

The great siren boomed, giving voice to the old Russian captain's desperation: he knew what his son-in-law was about.

The Zodiac rebounded off the *Vostok*'s hull, the impact crushing the light bar and throwing the Aldis from its mount, causing it to thrash around wildly, only to hit den Hartog in the face – breaking his nose.

The Dutchman knew he was done for. Consciousness was fading. His hand slipped from the wheel. The barrel, sailing through the air as though suspended by the arctic winds, hit. The Zodiac was no match for the tremendous force of the collision: the bow ruptured, the semirigid hull fragmented.

From her own Zodiac, Holding saw him. His head smashed against the Aldis lamp. His hand grabbed wildly, desperate to find the throttle that he would never find: it was smashed, ripped from its housing.

The barrel destroyed the Zodiac.

She thrashed her engine, the torque so massive that the bow rose eight feet out of the sea. She spun her Zodiac round and came at den Hartog's boat in a parabolic curve, closing at thirty knots, hammering her engine.

He was down. In the water. Her fault. Her obsession. She was captain. But she knew that she was better than this. That man on the *Vostok* was murderous. An animal, in love with killing.

She threw the throttle back, spun again, discharging her momentum against an incoming wave, and brought her boat to a standstill. The body. Where was the body? Then she saw him. She grabbed a guy rope, wrapped it round her wrist, and dived overboard. The shock was intense. Freezing cold. Unforgiving. Despite her training, all her experience, the freezing waters knocked the breath from her lungs. She surfaced, steam pouring from her mouth as she fought to spit out water and drag air into her burning lungs.

She saw him. His life belt had inflated, its fluorescent strips standing out from the inky water – a beacon. She kicked with all her strength, all her power. But her legs were useless, kicking against nothing, all sensitivity lost to a numbing cold.

Could it end like this?

Then she had him. Her hand found his chin and turned him round, so that he was floating on his back – his face free of the water. Her body was no longer responding – she couldn't even be sure that she still had the guy rope locked round her wrist. Holding felt sure that she was pulling with all her might. Heaving on her arm, levering against the tether, pulling herself towards her Zodiac, only three tantalising metres away. She was sure of it. Prayed that it be so. But her senses were not responding. Only when her face hit the rubber hull did she know for certain that they had a chance of survival.

She fought to pull herself aboard, but the height of the yawing Zodiac's bows seemed insurmountable. The rubberised hull was slippery, no friction; her numbed hands were next to useless. Den Hartog's life belt was lashed to her guy rope; it kept his head, which was obscured by a flood of crimson liquid, afloat.

One last effort.

The excruciating pain told her that she had ripped a pectoral muscle. At least she was alive. Bracing her legs against the hull, she heaved at Ronald den Hartog's dormant frame. Dead weight. Thirteen stone. She heaved, pulling one and a half times her own body weight. She screamed at him in rage, 'Move, you idiot! Move!'

He lay there, blue with the cold, hypothermic. No radial pulse – veins withdrawn beneath the skin. Or were her hands still too numb? She forced her left hand inside his wet suit, pushing the life belt aside as she reached for his heart. He was alive. And breathing.

The engine burst into life. She pulsed it, bringing the torque up, then let the blades churn the sea. Holding powered the boat away, peeling off to port, driving the small vessel's bows out of the water, planing against the imperfect surface of the oscillating sea. She could see the *Warrior*, standing out like a great haven of civilisation against the total darkness of the the cruel sea. The Zodiac banged into the surface, flew and hit again. Captain Alice Holding bit down, her teeth grinding, her jawbones locked together. Hatred, pure abhorrence, burnt in her eyes and her heart. She knew that Ronald den Hartog would live, and that she would find a way to make the Russian pirates pay for their actions.

Where was that damned coastguard ship?

Location: Ramada Inn, 48th Street and 8th Avenue, NY. Time: 14.50 (EST). Date: today.

He loved her. Made love to her. With her. He never thought he could be so close to someone, so at peace with total intimacy. More than just physically. So close emotionally, inseparable, to be one. Mark had loved Jenny for eight years. He never felt that he quite deserved to be so lucky. Many of his friends had to make do, or were still struggling, even at 28, to find the one person with whom they could be themselves.

He allowed himself to remember. So long ago now. University days – another life away, unrecognisable yet strangely familiar. They had played from the beginning. Jenny had just broken with her long-term lover. On reflection, it had been little more than a teenage infatuation, although at the time she had been deeply upset to lose her beau. Mark's past was still very much a part of his life at that time. His ex, Amy, was in the second year of a course in London, he a fresher in Kent. It was a long-distance relationship that was doomed to fail.

Jenny and he had felt the rush of excitement, the connection, even from the first moment they had met. Not love exactly, but a deep-rooted attraction that soon became a vital friendship. Two young people finding their way in a strange environment, both unhappy, struggling to find an identity of their own. Love grew. Not some half-understood emotion captured on the lid of a chocolate box. This was real love. Passionate love.

They had seemed terribly old-fashioned. Inseparable. They had picnicked by the stream, under the old oak, in the grounds that the university gardener wisely left to run wild. It was a personal paradise. A haven for pleasure, in an otherwise emotionally barren and single-minded environment. Their friends had laughed about their love affair, for it could only be described as such, and yet Jenny and Mark knew that they were the lucky ones. They had found something vital to life, but a something that was beyond the ken of their associates.

He loved her. He made love to her.

They had stolen horses from the university research stables at four in the morning. Jenny, a prodigious rider, and Mark, competent but out-classed, rode across the fields to Herne Bay. They had galloped along the sands, as the sun kissed the shoreline at the dawn of another day, spray drenching their clothes as they rode, the salt air burning at their lungs as they laughed incessantly like insane fools.

Empathically astute, she knew him, revelled in her knowledge of him. He loved that feeling. The certainty that, whatever he did, she was there, in his mind, in his heart. With him.

Their secret escapades were legendary, but this trip had been the best decision they had made in their time together. Eloping was an old-fashioned term for it, but what was so wrong with being old-fashioned? True, they had lived together all those years and so 'eloping' was not strictly the correct word – but no one knew, or even suspected. This marriage was for them. No one else. They had been alone, in America. Yet that was exactly what they wanted. They had always felt as though no one truly understood their relationship, the power of their feelings for one another.

He stared into her insatiable catlike eyes, the shocking-blue iris glowing in the half-light of the darkened room. She stared back at him, her lashes, so long and fine, catching the light as it spilt through a breach in the hurriedly drawn curtains, trimming her eye in a fringe of gold.

They loved.

Location: News International HQ, Wapping, London. Time: 23.50 (GMT). Date: today.

The landscape glowed like a carefully sculpted pile of embers. A glow that meant civilisation, society, technology. Humankind was incapable of seeing in the dark, so it employed its ingenuity and, in urban conglomerations, its most common form of habitation, it changed night into day. Street lamps radiated their phosphor orange energy, marking society's ephemeral foothold over the landscape. Elaborate architecture, another testament to humankind's status as dominator of Earth, traced the arc of the ancient River Thames. Tower Bridge, the easternmost bridge on the river, the great over-indulgence of the Victorian architects Jones and Wolfe Barry, stood like a fantastic make-believe structure belonging more to some never-never land than any real place. Yet there it was. Wannacheck traced the bridge's turrets and suspension wires with his eyes, following the great sweep of the tension-bearing rods, simple yet beautiful, like a butterfly resting for a moment over the river. So clever, so typically Victorian.

The breath of the beast whispered along the river, rose and curled itself around him. Wannacheck pulled his dark greatcoat tight, but knew that the cold he felt had little to do with the wind. He stared at the bridge and wondered about the great age of achievement that had produced such a fantastic structure. How naive society's development had been, how unabashed. With no regard to environmental concerns, no concept of ecology in the modern sense, simply the desire to further humankind's knowledge and economic might. He wondered if there would ever be an age like it again.

The Old Man had been more philosophical than he could ever have imagined. Their conversation had been the strangest of his life. More frightening, more real than anything he had ever known. He could see Jackson before him, as though he were standing there on the roof with him, at that very moment. He cogitated on their conversation, turning it over in his mind. That nauseous feeling ever constant, hitting him in waves. Was this how a soldier might feel before going into battle, knowing that he faced uncertainty and possibly even death?

'You don't believe the air-traffic-control strike, do you?' Jackson had asked.

'The Australian government's up in arms about it.'

'So they would have us believe.' Jackson searched Wannacheck's face for a reaction.

'It's a sham?'

'The whole bloody thing's a sham, Peter. We've been shat on good and proper. No one, and I mean no one, has any idea what's about to happen. Not even the kooks who've been burbling on about this on the Internet.' Jackson sucked hard on his Camel, holding the nicotine in his body for as long as he could tolerate. When he breathed out the smoke was barely noticeable, having almost disappeared within his system. He stared at Wannacheck, eying him keenly with hard, well-trained eyes. Wannacheck looked into the cold green pits for a sign that Jackson had some good news. There was none.

'Three weeks ago we had a call from a mole at Space Command in the US. They're the guys who monitor US airspace for abnormalities.'

'Watching for nukes and nosy neighbours.'

Jackson smiled, his yellow teeth a monument to his phenomenal habit. 'MI5 sent a man within the hour. He put a team from C-Section, protection and security, on twenty-four-hour duty.'

Wannacheck was incredulous. 'MI5?'

Jackson nodded. 'They've been with us ever since. Same down the corridor. The whole building's been locked down. Plenty of people coming in and out, but we've had friends with us wherever we've gone.'

'So how come I've been left alone?' Wannacheck watched the Old Man's face as it broke into a look of disgust. Not really directed at Wannacheck, just vented in his direction.

'Don't be so fucking naive. Your car, your phone, your flat. They even know how often you pop that little tart at Sentinel Online.'

'A D-notice, then?'

Jackson nodded, stood up – not too well, Wannacheck noticed – and walked to the window. He pushed the nicotine-stained fingers of his right hand through the office-grey Venetian blinds and looked out through the gap. 'A total gag. Not just us. Everyone. In the US, Europe, everywhere. None of us have been allowed to follow through with the story, never mind report it.'

He pulled open an old leather briefcase, a bit like a doctor's bag, and took out a bottle. Jackson had been on the wagon for four years – it was well known that his liver was so pickled that his doctors had offered to exhibit it at the Natural History Museum. Sclerosis aside, Jackson poured a generous helping of Jameson's Irish Whiskey into two dirty coffee mugs. Generous to the brim.

'Your a good man, Peter. You should've pushed harder, been less pompous. You should have stayed at the Beeb. Would've done all right.'

'Thanks, Dad.'

'I can't believe I'm spending my last night drinking this crap with you.'

'There must be something they can do?'

'There is: leave.'

Wannacheck looked at the images that the lab runner had brought him. The Prime Minister, his PPS, the Foreign Secretary and his new wife and most of the Cabinet. Thorbridge and his scientific colleagues from the DERA.

'They've left us to it?'

'Gone to the only place where they might be safe.'

'That's why there's a strike in Oz?'

'Got it in one. They're restricting entry to Australia without arousing suspicion. How better than to close the airports?'

'Christ. That's what Thorbridge meant: Australia is the Bulwark.'

'What did he tell you?'

'Just that I should take a holiday where it's warm in winter.'

Jackson made a raucous noise, like an old metal file grating against a rusting iron bar. Wannacheck had never seen him laugh before. Must be the whiskey. It was obvious that the current mugful was far from his first drink that evening.

'Couldn't if you wanted to. He always was a twisted bastard. Met him in sixty-nine when he first came across. Story went that he killed the checkpoint guards with piano wire.'

Wannacheck ignored him, determined to follow his train of thought before the alcohol finished off his already numb brain. 'So the Astra satellite losing a solar panel wasn't an accident?'

'Mike, on the science desk, reckons it was a piece of debris that hit it, ahead of the larger body.'

Wannacheck stood up and walked to the window, turned and studied Jackson's eyes. His pupils were dilated – too much alcohol. Small pearls had welled there. Could the Old Man be crying, or was it just the smoke from his Camel cigarettes?

Wannacheck turned away as he realised Jackson was looking at the picture hanging on the wall. It had always been Jackson's only concession to his private life: a photograph of his two teenage daughters.

'How long?' Wannacheck asked.

'They wouldn't tell us. In case we found a way to get it out there.' Jackson looked at Wannacheck, full in the face, his jaw muscles taut with anger and pride. 'They were right to shut their mouths. We would've done it somehow.'

'Don't you have any idea?'

Jackson sat down, took another drag on his dying cigarette, looked at the clock on the wall, and said, 'Mike reckons it's soon. Judging by your pictures I'd say he was right.'

'Tonight?'

'More likely tomorrow. It'll take the great and the good quite a while to reach Australasia.'

'Tomorrow, then.'

The fingers of the beast were at his throat – he could feel it trying to tear through his clothes, breaking in at the seams. He shivered, involuntarily. He was afraid. Not of death itself – he could not even imagine that. No, he was afraid of the uncertainty, the uncharted waters. If only he had some idea of what to expect, at least then he could prepare himself. That was the cruellest part of all. The government could have at least given its people some warning so that they

184

could make their peace with their loved ones and their God. But maybe they were right to stay silent. What would have been the outcome? Mass panic, suicide, riots? Maybe even war, if Europeans had tried to flood off the continent across into Asia, perhaps via Turkey. Maybe they had taken the only option. What nation on Earth was large enough, which economy robust enough, to house half a billion people? None.

He knew he should go to her. But he did not love her, nor she him. Amanda was a good woman, but he could not face her now. Did not even know if 'they', the secret-intelligence operatives, would let him go. No, he would face what was to come on his own. The roof was a great place to do that. At least from that height he might see what was coming before it hit him. It was the only way he understood. He did not know how he would react when the time came. Could not stand the thought that he would break down, that his nerve would desert him, when he stared death in the face.

Location: the North Atlantic, 150 miles off Land's End, southwest Britain. Latitude: 45 degrees 5 minutes north; longitude: 5 degrees 25 minutes west. Time: 07.32 (GMT). Date: tomorrow.
The scream was enormous, godlike. It was the air itself, screeching as it was torn apart by three burning globes: massive, glowing, rotating, burning like hot coals spat from some unfathomably large fire. Smoke, flame and debris traced their path across the crimson morning sky, fragments of an old comet broken up by the heat of the sun, as they encroached on the inner solar system. Then, captured by the gravity of the third planet from the sun, they began an inexorable journey towards its surface. Towards Earth. Each burning globe was impossibly huge, weighing 200 million tons, measuring one-third of a kilometre in diameter. Less than 100 metres apart, they tore their way into our atmosphere, entering at a 45-degree angle to the planet's surface, losing mass and energy as their faces were burnt by friction. It was almost as though the air itself were a hapless child attempting to blow away some terrible evil.

The blocks of ice and rock smashed through the North Atlantic seas. Accelerated by Earth's gravitational pull, they hit ground zero at forty miles a second. Searing, blinding energy. Beyond understanding, beyond even humankind's ability to destroy: a burst of energy so enormous that it was equivalent to 300 gigatons of TNT, ten times the size of the world's nuclear arsenal at the height of the Cold War;

it vaporised 400 cubic miles of earth, rock and sea. A monstrous wave front, a shockwave, burst away from the impact site, rupturing the atmosphere. The semidarkness of morning was banished as an incredible radiation of light, a detonation flash, thousands of times brighter than the sun, turned night into brilliant day for most of the northern hemisphere.

Time after impact: +8 seconds
Millions of tonnes of ejecta, cometary and other debris, were hurled into the air – at 1,000 kilometres per hour. The ejecta raced away from the planet, fragments of comet flung back into space for the last time, intermingled with water vapour and earth, becoming a moving wall of gas and rock – a great mass of detritus, superheated by air friction to temperatures in excess of 3,000 degrees Celsius.

Time after impact: +43 seconds
Ripples, great wave fronts racing away from the impact site. Not water, but the Earth itself. The pressures released were so high that the land became like water; its great front, hundreds of miles across, raced towards its target: the British and European shorelines.

The beautiful green land of Ireland was swept aside, its shoreline, a weak barrier for such an impossible force, rupturing and buckling like bubbling liquid, thrown on to the crest of the wave front, carried along on the shockwave's inexorable journey. Trees, houses, rivers, all landmarks, all solidity, lost. No earthquake, no tremor, could compare to the devastation wrought that day.

The land, humanity's only constant, not only moved but was utterly changed, thrown out of all recognisable pattern and behaviour into a totally new and alien form. The concept of emergency response was irrelevant. There was no Ireland left to respond. Plants, animals, humans, all dead. Even birds escaping on the wing were robbed of their lives as the shockwave generated great downdraughts and updraughts and air-pressure fluctuations that forbade any creature to escape. The Irish Sea was destroyed as the wave front raced across the continental shelf, churning the land beneath the sea and throwing it into the wave.

The great radiating death hit the southwest coast of Britain, carrying the pulverised land mass of Ireland on to the mainland.

Time after impact: +10 minutes
The molten debris, the raging melt sheet thrown away from the Earth at the moment of impact, re-entered the atmosphere at near-ballistic velocity, like a million missiles.

Location: Ramada Inn, 48th Street and 8th Avenue, NY.

Innocents slept, their leaders in absentia, as the burning fragments traced their path back through the atmosphere. Long before the destructive wave front of the great impact shockwave toppled the great buildings of the East Coast of America, the melt sheet terminated millions of lives.

Mark and Jenny had loved. They knew how to live their lives together; after eight years of love and hardship, joy and achievement, they knew. Yet they had never given a moment to think how they might die together. That day, their love was terminated.

A great shower of death, travelling as fast as a bullet from a high-velocity rifle, lit up the night sky over New York. First the great impact flash had stirred most of the sixteen millions, many of them fearing that nuclear war had finally begun. Within minutes the falling death had arrived. People did not know how to react. What should they do?

Columbas Circle and Carnegie Hall were the first to be destroyed, their façades, so carefully conceived, constructed and cared for, ruptured by the melt sheet. The falling death crept along Eighth Avenue, to its intersection with 48th Street. They died, Mark and Jenny, together, holding each other. Others died afraid, lonely. Some, still asleep, died in merciful ignorance as the debris shattered their apartment buildings, or the great fireball, started in Midtown, ripped through the neat grid-plan structures of New York, like some natural wildfire through a forest of dead trees. Millions of people, dead. No warning, no escape, no hope.

The emergency services were not ready – they had had no warning. A moratorium, as in other countries, existed to protect the innocent from panic, looting and public unrest. They responded to the emergency, but many of the brave participants were killed simply trying to travel through the chaos that had once been New York city streets. They tried to cope. To restore order. But they did not know what they were facing.

The shockwave raced through the darkness, having already passed from day into night, eating ocean, land and air, as its wave front moved inexorably towards the Northeastern Seaboard of the Americas.

Location: News International HQ, Wapping, London.
He had seen the flash. Known that it was coming. He no longer felt the cold: he was beyond such feeling after his vigil. He had waited, watched and had been rewarded. He had seen death. He knew it was close now. Wannacheck walked to the barrier surrounding the roof of News International, 120 feet above the ground. He watched the morning sky, dark-blue and purple, as all November mornings were wont to be, turn slowly into an unnatural red.

The sirens and bells were a surprise: someone obviously felt that the bomb had finally been exploded, felt a duty to alarm everyone. It was almost deafening, wailing, screeching. He shuddered. He could hear the sound of police vehicles and ambulances, fire engines and even army trucks, racing through the streets. The roads were jammed. From his place of vigil, he could see the congested roads, a solid stream of traffic, attempting to cross the Thames in both directions. Where were they all heading? Emergency switchboards must be jammed. The number of emergency vehicles on the streets was astounding.

Fire, rock and water, like some terrible biblical retribution for humankind's evil, fell from the skies.

The people of Britain turned to their televisions and radios, long seen as the most reliable source of wisdom, for salvation and advice, only to find static. The colossal energy released at impact had flooded the atmosphere with charged particles. The country, soon followed by mainland Europe, was deaf, dumb and blind: its radio, telecommunications and radar either destroyed or rendered useless against the falling melt sheet. Many of Europe's telecommunications, military and environmental satellites were now raining down on Earth, as they were destroyed when the atmosphere ruptured and the melt sheet burst away from the impact site.

Oxford Circus, one of the central hubs of London's underground train system, was very busy at 7.43 in the morning. Amanda knew something was wrong as soon as she stepped out of the strange cocooned domain of a Central Line tube train. The platform was always busy, a constant throng of commuters and tourists, even in November, but she had never seen it filled with hundreds of people,

188

many of them burnt and bleeding, most of them screaming. The platform's air, ever stale and thick with the odour of other people, was rank with the stench of charred flesh and smoke, which was evidently pouring into the underground tunnels from the surface. The smoke clouds licked at and obscured the dirty curved ceilings and electronic signs suspended over the platform. The lights, ordinarily a constant brilliant white, flickered like dancing flames.

Fire. Her first thoughts were of a fire. A number of incidents in the tube system, notably a blaze at another main hub of the system, King's Cross, had resulted in terrible injuries and death. She worried for her sons, her boys. What would they do if she were gone?

The platform shook as an explosion ripped through the subterranean network, the result of more debris bombarding the surface, destroying the Oxford Street façade at the intersection with the sweep of Regent Street. The great mass of falling ejecta turned London into a raging fireball, worse than even the Great Fire of 1666 itself.

The explosive boom amplified to a bass toll, thanks to the action of the interconnecting tunnels, which functioned like echo chambers. The lights failed. The screaming was deafening. Especially her own.

Wannacheck saw the wave front. Even through the heat haze, the soot and falling ash of the fireball that was London, he saw its approach. From a distance of five miles he could see the oscillating arc of destruction as the great wave front, now somewhat dissipated but still powerful, shook the very Earth itself in a creeping line of destruction. Now he knew how it would end: he had waited all night for the certainty, but now it was here. He felt the first tremor, the first quake, as it travelled through the roof beneath him and along his spine. He knew how he would die; he could see it. Tower Bridge, the great Victorian icon, shattered, crumbled like a dead leaf. He saw the perimeter wall, the great fortification that had given News International the name of Fortress Wapping, collapse. He was thrown from his feet. The building vibrating, shaking wildly. The roof collapsed. He fell, screaming, until he hit the floor below – but didn't hear the sound of the bone and cartilage in his neck, yielding under his body, generating a sickening crack. He was silent, but the earth was not. Its voice was deafening, violent, an incredible din: the sound of earth tremors, quakes changing the land beyond all recognition, destroying thousands of years of civilisation. London was dying.

189

Time after impact: +15 minutes

A wall of undulating, supple destruction, hundreds of feet high. It raced in concentric rings away from the impact site. The great waves, triggered by the massive forces of the explosion and the resulting submarine landslides on the continental shelf, raced towards what remained of the British Isles – already greatly changed.

The power and speed of the water, this great tsunami, simply overwhelmed the land. Any structures left undamaged by the great pressure front and the falling melt sheet succumbed. Any tall buildings left standing after the shockwave succumbed to pelting from the falling incendiary sheet of ejecta. The wave raced across the surface of the continental shelf, shattering the ravaged coastline, flooding inland, swamping the once green and lush landscape to the plains of Salisbury, eastward towards the forests of Hampshire. The western coastland, already crushed by the brunt of the shockwave, was lost to the great flood – its incredible ferocity, size and sheer power consuming all that stood before it.

London was a fiery death zone. Smoke climbed to the heavens, blanketing the metropolis in soot and ash – a combination of wood, fabric, plastics and rubber. The air was almost unbreathable, poisoned by burnt fuel and chemicals spewing from refineries along the Thames, and from factories and fuel stations all over the city. But the foulness of the stench of blazing industry was easily exceeded by the terrible, dark reek of burning flesh and bone.

Landmarks all but gone, any survivor would have been lost in the blazing rubble. Even Canary Wharf, the embodiment of modernity, had crumpled, its great tower shattered, not even a stump left to remind survivors of its former mass. St Paul's Cathedral, perhaps bolstered by its hemispherical structure, appeared to have survived, at least partially, intact – though, had any observer been able to look more closely, it would have been obvious that most of the structure had collapsed: even while the great dome survived, much of its foundations had crumbled.

The Thames Barrier, a gesture to the acknowledged fact that London had long faced the probability of succumbing to a great flood, stood idly by. Its great shields were largely shattered, thrown free of their delivery mechanism by the shockwave, and protruded from the bed of the Thames like giant scallop shells.

When the great flood, feared and prepared for since the barrier's opening in 1982, arrived, no one was left alive to activate the metal

190

sentinel. The barrier was useless – wrecked just like the city it was meant to protect.

She could hear the roar. Like her fellow passengers, Amanda was trapped in a darkened hell. Smoke had filled the cavernous tube network, pouring down from the streets above – as the fireball spread across the capital so the underground network had become a deathtrap. Many people had taken refuge on the platforms, believing, as their forebears had when Adolf Hitler's Luftwaffe had bombed the city, that the tube might offer them protection from the falling death. They were right. The ejecta wreaked havoc on the surface but left the tube largely intact. The ensuing fireball, however, eating all combustible fuel on the surface, had poisoned the air above and below ground.

Amanda, knee deep in water, screamed as an object which blocked her path turned over. She could not see it in the virtual darkness, but instinctively she knew that she had touched death. A cold and rubbery shape. Two small, spongy hemispheres and a squat protuberance, small and firm, yet misshapen: the face of a dead child. A child drowned, trampled in the madness of panic, or crushed by falling masonry when the shockwave had violated the great sunken walls of the underground network.

The roar came again. It was different from the other sounds she had heard. Despite the injuries to her head and legs, she was alert. Adrenaline had focused her mind, making her keen. It was water – the sound of water. She had already come to understand the nature of the water that she stood in – a leak from some collapsed retaining wall in a far distant tunnel under the Thames. No, this was much greater.

Her head reeled with fear, her panicked heart pounding in her chest, making her desperate to take action. But what? Her throat was throbbing, sore from screams of frustrated rage and the choking air. She tried to find the stairs in the darkness, but her progress was hampered by others on the platform, many of them angry and frightened. She moved gingerly, her face already badly bruised. A man with powerful hands had grabbed at her face, not realising his mistake, and had used her for leverage, pushing her head violently, sending it ricocheting off a wall. She had sunk back into the water, senseless, sickened.

Amanda stopped and listened. She was not alone. Everyone heard it – they were silent. Their endless groaning and mutterings, screams and curses, simply stopped. Her heart raced ever faster. She was hyperventilating – shock and fear combined with smoke inhalation. It was now – she could feel it – she was going to die now.

The great liquid claw poured through the caverns of the London underground system. Millions of gallons of water, having already swamped the surface, poured through the restricted openings of the wrecked station entrances, finding the path of least resistance. The near-cylindrical walls restricted the flow and so greatly increased the pressure levels; water poured through the underground network – exerting thousands of pounds per square inch of pressure.

They screamed, but she could only stand and listen. Her voice gone, energy drained, her hopes dashed. Amanda thought of her boys, no doubt dead, somewhere on the surface, near their school in what had been Twickenham. But she could not see them that way: they were vibrant and wonderful – so very full of life.

She died, knocked from her feet, swept along by the tirade of water, killed instantly as the incredible pressure of the water rammed her against the crumbling walls of a place once so reassuringly familiar and now so deadly.

Location: North Sea, 20 miles SE of the Shetland Isles.

They resented Alice Holding, the crew, all of them. They had all served with Ronald den Hartog. They liked him, or liked to think that they did. She knew it. As captain she had to take responsibility. She stared at the cabin's reading light, almost directly above her head, until it hurt. Her berth was dark. Her face was agonising – burnt as she had fought with her crew to extinguish the fires on deck. The melt-sheet debris had laid its destructive hand even on the *Warrior*. She knew something had happened, something bad. No radio contact, no GPS. Nothing.

She sighed, pushed her head more firmly into her pillow and stared up at the reading light above her head; it glared at her, accusingly. 'They' still resented her for staying with the *Petrushka*, even until she had been escorted to harbour by HMS *Belrophen*. Maybe she should have turned the *Warrior* round and headed for port, but den Hartog had given his all to stop that ship. She had other commitments apart from personal safety. They all did – risk was part of their lives. They were scared – she could smell it. They were scared that something terrible was happening. She had to concede that she shared their fears. She had never seen the magnetic compass go wild before. Never. She could cope with the uncertainty of a mission, but the uncertainty of location? Without GPS and a reliable magnetic fix they were in trouble. The Decker navigator had been useless from the moment the falling incendiary fragments had destroyed the shore-based radio transmitters. The ship was in deep trouble. She knew they were glad. They wanted her to fail. She knew it. For the first time in her life Alice Holding knew the fear that comes with failure.

The wave, 200 feet high, came from nowhere. No warning. No chance.

The *Rainbow Warrior III*, a ship designed to fight for the Earth, its ecology, the safety and health of its animals and plants, died as its crew died, tossed like a spinning top, as the wave hit the ship broadside on. The *Warrior* upended and went under, carried at the zenith of the wave – driven into the ocean like a nail into wood. Her mission, their mission, irrelevant, over. Terminated.

The wave continued. The Netherlands was the first nation on main-land Europe to succumb. The flood, biblical in its unremitting ferocity and scale, swamped the lowlands, extinguishing the fires but destroying all in its path. Worse than any flood but Noah's, the great waters lashed Europe's shores.

A great fireball, started by the falling debris, had already turned the Benelux countries into an inferno. The wall of fire, dampened only where the tsunami smothered the land, crept steadily over the continent, highland and lowland, the great seat of civilisation, ablaze. All that was combustible, all that was living and most that was not, burnt.

Panic reigned as the people fled for the hills. Murders were committed as neighbour fought neighbour, clamouring to reach sanctuary. In Austria, the people knew almost instinctively that they must reach the ancient symbol of constancy and sanctuary: the Alps.

TWO

AEONS OF FIRE AND ICE

Time after impact: +3 hours
Orange, carnelian, heliotrope, bubbling glowing yellows, shot through with iridescent ribbons and jets of gold and purple-blue. The fire swept all things aside. So intense was the heat generated, over so large an area, that the firestorm became self-perpetuating, like the worst stories from bombing raids over cities during World War Two. This natural tongue of hell licked across the European continent, ingesting, destroying, in 2,000 degrees of searing heat. Dante's Hell on Earth.

Thousands of columns of hot air carried sheets of smoke, soot and ash into a sky already perturbed and darkened by slow-falling ejecta debris and dust.

Location: Cogema Beach, Northern France.
The seas were restless, grey, crashing hammers of water, lashing the once beautiful shoreline that was now reduced to a grey murk covered in soot and ash, its flora and fauna charred, crackling and snapping in the intense heat. Above the beach, the plant. It had stood, apparently resilient to all tremors, violence and turmoil since the impact.

Jacques Durrant had stayed. How could he leave when the plant itself represented such a great threat? His colleagues had left, gone to find their loved ones – probably to die with them. He knew that his career was at an end in any case. He was old, felt old, yet his friends all said that 58 was too young an age to retire; his doctors disagreed: it was not too young if he was to enjoy any form of retirement before the end.

He knew it was hot. The external instrument stack, located on the roof, had given him a lot of valuable data, until it failed. The computer had given him a reading of 190 degrees Celsius outside fifteen minutes ago. He could not feel the heat at first: the air-conditioning plant was very efficient – had to be.

He had always wanted to see Africa, especially Zimbabwe, to travel with Clarissa and experience the Zambezi. They had dreamt of making love under the pounding waters of Victoria Falls. They had always indulged their fantasies, but that one had escaped them. They had built the idea into a great ambition that must be fulfilled, one day never.

Darkness, light, darkness. He knew what the power fluctuation meant. He cursed the suit: he needed to wipe his brow – the sweat was stinging his eyes, blurring his vision. He had to see, damn it. He told himself that it was his nerves making him perspire. He knew it was a lie, but he thought that the lie might just hold long enough for him to cope without screaming.

Of course he knew that Clarissa had found out. She had known him too long to miss the signs, but she had been wonderfully secretive about it, helping him to ignore the reality. It was the only way he could cope – she knew that, could read him like a book.

His blue Mont Blanc pen, a birthday present last May, rolled from the desk and hit the floor. It shattered, throwing a great blot of black ink across the floor. Jacques stared at it, a grotesque mess, just like the lump in his lungs, spreading, black, abhorrent.

The desk oscillated, his control panel shaking in its cradle. He could not ignore the vibrations any longer. The tremors were getting worse, or was he losing his sense of balance? The heat, must be the heat. The shutdown was simply not responding as per the practice drills and simulations. The bank of control monitors were blinking in symphony with the ceiling-mounted strip lighting, the main computer crashing and rebooting. The control systems, without a computer, were beyond the influence of one man, or even a dozen. The systems were, like all electronic components, designed to operate within specific temperature tolerances, usually 35 degrees Celsius. He knew it had to be 80 degrees inside the complex – he had experienced heat on holiday in the desert, but nothing like the excesses he was suffering now.

The fail-safes might yet work. He had no way of knowing. He had tried, time and again over the last hour and a half, to raise outside assistance: the emergency telephone, radio and email systems were all off line. He sank back into his chair, the air-inflated rubber bodysuit crumpling and buckling around him like a strange extra skin. He stared at the cylindrical control room. It was smashed, the instruments dead, the very walls shaking. He had stayed, like the captain of a sinking ship. What else was the director to do?

He thought of the water, falling, flowing over the famous precipice, the majesty and power, as unchanged as the day David Livingstone had discovered it.

La Hague power station became the 52nd of France's nuclear power plants to explode. They had all been built to exacting homogeneous designs, designs that had never incorporated the concept of immense heat or impact with falling ejecta. No one was alive to hear the wrench of concrete, the crack of potential, turned to kinetic, energy. Nor was anyone able to witness the great pall of superheated irradiated steam as it climbed into the atmosphere.

Cogema Beach was a scar. A deadly hole, littered with detritus of concrete, steel, graphite and uranium 235.

Location: ice cave, Eisriesenwelt, Austrian Alps.

Time after impact: +6 hours
At least it was safe. They just knew it, sensed it. Dark, cold, a total contrast with the fearsome sight they had left outside. There was a certainty here. Even though ice had fallen from the walls during the aftershocks, they felt certain that no wall of flame could reach them here, in this frozen kingdom.

It had been an awful sight to behold. An experience no human being was meant to face. Helen, Jack and Iain had reached the proximity of the caves with ease, as their climb had covered much of the difficult terrain already. They had intended to descend the next day, then disaster came to the Alps.

They had seen it: an immense flash that woke them, illuminating their tents, bursting through the canvas as though a thousand searchlights had been thrown on to their location. Just as quickly, the light had gone. When they looked out of their tents, they saw an early-morning sky like no other. Iridescent, eerie, though not beautiful in any way: it was crimson, blood-red. No shepherd's delight, it brought only terror.

Iain had reacted first. Jack had to concede that Helen had been right to bring the Troll along: he was wily; he had literally saved them. The tent had been pitched in a restricted area atop the mountains. It had been a cold night, but they didn't care. After all, they were students and rule bending was almost part of the job descrip-

196

tion. They were close to one of Austria's most famous sites. They had intended to take it in before descending back to sea level.

Iain had ushered them out of the tent, grabbed their kit, or as much of it as they could muster in thirty seconds. They dragged their clothes and boots on and were gone. The tent stood alone, abandoned, left to watch the creeping death as it spread across the land, heading inexorably for the Alps.

Fifteen minutes after the light burst, the shockwaves had hit. Iain was a rock: even when Jack's knee buckled under him, he simply picked him up, carried him across his shoulders. The land buckled; even the great folded might of the range heaved under the pounding. The three were tossed like feathers as the great mass of rock and earth heaved under the shockwave. The sound was almost unimaginable. Even though the brutal force of the impact had dissipated somewhat, the wave made the range scream in protest.

They heard but could not see the landslide. It was often the way in the mountains: sounds bombard you as though their originators were at hand, and yet they were kilometres away.

They knew that they were in real danger when the fires came. Jack had been first to see them, as they burst forth as if some great flame-thrower had spit its death over Europe. Helen took charge, leading their descent to the tourist entrance of the caves. Her instincts were second to none.

Jack had refused at first, but soon realised that Iain had the answer. If they waited for him to half walk, half climb, they would be dead. He had protested, told Iain to get Helen down to the caves. Iain simply told him to shut up and ignored him.

Hefting Jack, no waif, on to his shoulders – laying him across his back like a medieval hunter carrying the king's dear – Iain climbed down the steep screes and craggy cliffs, all the while nursing his passenger. It was the damnedest display of bravery and strength that Jack had ever witnessed.

Helen's brilliance shone – she had led them to the well-trodden path within seventeen minutes. But there was no safety there – all hell had broken loose.

Just as the flames crept across Europe, taking Austria in their wake, the people had fled. Maybe Jack had done something right. It had been his initiative to take to the Alps in the first place – his rite of passage. Whereas yesterday they had been all but alone on the great mountain, today the great edifice was alive. Seen by the gods, it could easily have been mistaken for an ant hill, but these ants walked upright, bipedal. Children clung to the breast, exhausted parents half dragging, half carrying, all that they had salvaged. Some were more experienced than others, fit, stealthy, carrying adequate provisions in well-designed rucksacks. Others had obviously escaped with their lives and little else.

'Some of those kids are naked,' Helen said. It was barely a whisper, but they heard her. Iain set Jack down, offering him a shoulder to lean on as Jack put his weight on his right leg – the left knee was useless, maybe dislocated. Jack locked his arm round his saviour's thick neck, never more pleased that the Troll had a reputation for being made of the rocks he so loved to climb.

He could see the fear in Helen's eyes, could feel Iain's response: his heart pounded hard – not just because of his titanic efforts with Jack – he was too fit for that to trouble him. No, he was afraid.

Jack looked Helen in the eyes. 'At least they made it out before the town got it,' he said, nodding in the direction of Tenneck. She tore her eyes away from the desperate throng that flowed over the meandering path: a river of human flotsam. Then he saw it in her eyes, the look that must have greeted her bastard of a father when he beat her. Terror, agony, grief. He saw all the world in those deep-green eyes. Her cheek, ashen pale, rose to meet her lashes as she winced at the burning tear that spilt from her eye. Defeat. It was terrible defeat that he saw there.

He was damned if they would give up now.

'The fire will be here, soon. They –' he nodded at the people as they struggled up the mountain path '– will be here in minutes.'

'Whatever's caused this is not done yet. Look.' Iain pointed to the southwest, thousands of feet below; the great body of water that gave Tenneck its beauty oscillated, tracing a great wave front across its surface. 'Aftershock again,' he muttered.

Jack became aware of the noise for the first time. Maybe it had been the pain or perhaps the shock, but he had been totally unaware of the absolute din that reverberated about them. Voices, thousands of voices, wails and moans, children screaming; it was hellish. Behind them the entrance acted like the sound box on a guitar, giving resonance and bass to the noise of the Austrians as they filed into the great cave mouth.

They were surrounded by a sea of people. Hundreds queued to get into the great protective might of the mountain. Austrian efficiency had ensured that the doorway was ridiculously small and resembled a domestic door. He could not help but check himself as his mind raced back to childhood stories, wondering if he would find a roaring fire in the hearth of some bizarre Victorian parlour.

'God no!' Helen whispered, snapping him back to reality. She had seen it happen: now the sound followed the sight, a bass roar, tearing up the rock face like a dragon gone wild. Jack heard the town of Tenneck explode. They screamed. Any calm established because of the physical effort to trudge up the tourist path to the caves evaporated as they heard their town expire. The mass of escapees burst into a cacophony. Panic, this was panic.

Iain looked straight into Jack's eyes, they both knew, in that brief exchange, that they would die if they were to hesitate for a moment longer.

Steadying Jack as he hobbled to Helen, Iain glanced back at the doorway behind them. It was madness. The tiny opening, wedged in the great rock face, restricted people's entry. The almost orderly queue that had formed was orderly no longer. Squabbling, thrashing, desperate, the people fought for entry.

Jack's hand locked tight on Helen's arm. He squeezed almost too hard, forcing her to look at him, to register their predicament. 'It's now or never, Helen.' She closed her eyes, turned and stared at the cave mouth, almost as if she had not seen it before that moment. He saw her hairline tighten, as she bit down, demonstrating, not for the first time, a gritty determination that Jack had come to love. It was a determination to survive born of betrayal. She took his weight, signalling to Iain that he should let go. The Troll was uncertain, but Jack nodded, sure that he could take his weight with Helen's help. Then, with a nod of Jack's head, Iain went in. Just as though he were on the rugby field, he used his massive shoulders and meat-hook hands to clear their way.

They were through the strange anomalous door and into another world. Frozen, dark, eternal in a way that a landscape changed by seasons could never be. They were in the world of the Ice Giants – 25 miles of eternal ice caves – safe from the terrors outside.

Location: Canberra, Australian Capital Territory.

Time after impact: +20 *hours*
'... FIRED ON PROTESTORS AS THEY CLASHED WITH POLICE. VISITING BRITISH MP DAVID GRIFFIN SAID THAT

HIS COLLEAGUES WERE UNRUFFLED. THE BRITISH PRIME MINISTER IS EXPECTED TO CALL FOR A UN EMERGENCY ASSEMBLY IN CANBERRA THIS AFTERNOON, TO DISCUSS RELIEF AID. BUT FEARS PERSIST, DESPITE THE US AMBASSADOR TO AUSTRALIA'S ASSURANCES, THAT THE USA IS IN NO CONDITION TO FULFIL ITS . . .'

He felt like a cheat, was a cheat, yet this man made him feel worse somehow. He made him nervous. 'I have a speech to prepare,' he said, dismissing the man.

'Make it a good one,' Thorbridge said as he rose to leave. 'This rests on your shoulders. Yours and ours.'

'Ours?'

'We should have tried harder to convince you.'

The British Prime Minister looked up at the big man. Those damned eyes were on him again. They were insufferable, just like their owner. 'Damn it, man, I've got guilt enough for both of us.'

Thorbridge smiled. He had dismissed the man as an idiot long ago and the Prime Minister knew it. He turned away, walking past, and almost over, the Prime Minister's private secretary, Sir Charles Price, to stare out of the window. The light was weak, a strange orange glow evident even here. He calculated the enormity of the impact if such vast quantities of dust were already being carried around the Earth to Australasia. He leant against the window frame, the tiny panes of glass distorting the view of the protestors who had appeared in the street outside the Parliament building. They blurred and merged to become one large streak of colour and movement against the sterile streets of the Australian Capital Territory. Thorbridge marvelled at their rapidity of response. He noted CND and Greenpeace banners among their number – no specific signs, simply any item of protest that they could find. No one likes to be left in ignorance, except politicians, he mused.

Thorbridge wondered about Europe, the people who had died there. He could feel little grief, but he did wonder at the scale of loss, the culture and learning. At least that stirred his emotions. Not grief: anger. He abhorred waste. 'Not enough guilt by far,' he said.

The Prime Minister looked away from Thorbridge and inclined his head. His personal private secretary looked into the Prime Minister's hazel eyes, eyes that had secured his election because of the sincerity factor, as the spin doctors had called it. He saw that Thorbridge was hurting his leader. He looked witheringly at the windows, making eye contact, but the large head dismissed the inferior with an arrogant twist, expressing his utter disgust. Thorbridge left the room, slamming the door behind him.

'I don't want to see him again, Charles. He's of no use to us now.'

The Prime Minister's friend and colleague nodded in acknowledgement, but could find little comfort in the news he was to pass on to the Prime Minister. 'Pakistan has launched a nuclear strike against India. Three missiles, warhead megatonnage unknown.'

'I thought Space Command's network was out of action.'

Sir Charles Price nodded. 'We found out the old-fashioned way.'

'A man on the inside?'

'Right. Satellite phone. Evidently not all of our satellite systems are down.'

'Just the ones that matter.'

The firebirds raced across the ancient landscape, its culture so old and remarkable that none save those who originated there could truly embrace its magnificence. Yet it was a culture taut with conflict. They traced an arc through the fog of water vapour, smoke and ash carried on the winds. Rain pelted their metal hides, running along their lengths and evaporating the instant the water reached the rocket motors. Ex-Soviet ICBMs, deadly, rarely fired other than in tests, these cumbersome relics of an era long since past were never fired at their intended enemy, the USA. Now they would bring even more death to a world awash with endings. Meeting no resistance, the emotionless killers raced across the Rann of Kutch, over the border and into India.

'What will that do to the area?'

Charles Price stared at his youthful Prime Minister and wondered if this man truly had the backbone he pretended to have. 'Total destruction for a maximum of a twenty-five-mile radius, fallout over at least one hundred miles. They evidently misinterpreted the impact symptoms. Thought India had launched a strike.'

The young man stared at Price. He was an old bastard, the records said so, but he was a good man in a crisis. With a stronger stomach than his own at any rate. He wondered if he should put so much faith in his advice. Too late to question his decisions now. 'I did care once, you know.'

'Yes. We all did.'

* * *

At 22 degrees 36 minutes north, 88 degrees 24 minutes east, Death walked the Earth again. Humankind proved to be his tool this time. Not to be outdone by nature, humanity's devices destroyed where life had struggled to survive. The mighty daggers ruptured the heart of Calcutta. All that had survived the worst catastrophe in human history, protected from the shock of the impact by the great backbone of the Himalayas, died. Humankind's worst enemies, ignorance and intrigue, had cost it another enclave of its beleaguered species.

Location: Kiev, Ukraine.
'Channel 19. Piper Cub Charlie 9194 calling. Over. Anyone receiving? Over. Piper Cub Charlie 9194 calling. Channel 19. Over.' He wiped his mouth, could still feel the stuff on his lips. It made them slightly numb. Couldn't pronounce its name – all the cheap vodkas were the same: bad names, bad hangovers. At least they were cheap.

The plane's Lycoming four-cylinder hummed a little too loudly as he tipped the nose up, pushing the ageing but ever-serviceable Piper Cub into a climb. He eased off on the yoke as he reached 2,000 feet. Holding the yoke with his left hand, he leant across the cabin and plucked the open bottle from the empty copilot seat. The rim irritated his lower lip, grating against the ginger stubble, a little residual alcohol burning at the roots of his day-old beard. Not caring about the instruments, or the view, he took a swig, gulping as much of the burning liquid as he could stand without gagging. Fire – his throat was on fire, just like the city he had come to regard as home.

'Guess no one wants to talk. Can't say's as I blame you.' He spoke into the open microphone with all the care and attention to regulations that one would expect from a drunk. 'If there's someone listening I need to let you know what I see, which ain't much.'

He pushed down on the right rudder pedal, eased the stick over, and banked. The Piper banked hard, giving Hammerick Bryan, its pilot, a good view of the landscape beneath. Black, everything was black. The world had gone mad, he felt sure – or *he* had.

'Everything's black and white, like one of those war photographs, just smoke piling into the sky. It's so thick up here you could get out and take a walk. I know this, though. Nothing new to me. It's war, this, it has to be. I know it, seen it all before. But Jeez! This is a bad one.' He released the hot key on the radio, waiting for a response.

Nothing. Silence.

He threw the bottle back on the copilot's seat, spilling much of what little remained on to the well-worn leather upholstery. The little old single-engine plane circled on over Kiev. He took Charlie 9194 down to 500 feet as his course brought him over the Dnepr. He levelled off and followed the river, taking him into the heart of the city. Then he banked off over Darnitsa, on the left bank. All was ablaze. At that level he could clearly distinguish the smouldering fires, still burning in the low-level boxes that passed for industrial warehouses and factories. Two huge warehouses seemed to be still ablaze, while a vast unmistakable pile of burning rubber from the tyre factory created two thick black bulbous columns of smoke which climbed skyward. They were so solid that they seemed almost as though they were the legs of an ebony god, standing proud over his destructive day's work.

He toggled the radio hot key. 'Darnitsa's finished. Pity. I liked those guys at the tyre factory. Only people I could still get tyres from for this old bird of mine.' He flew between the god's legs, his nostrils filled with the stench of burning rubber, polycyclic aromatic hydrocarbons filling his lungs with toxic benzenes. He breathed deeply and laughed, half choking, half cackling like a mad hen. 'Phew, making quite a stink up here, guys!' he yelled through the window, as though his factory-working friends could hear him. 'Guess we won't be on for poker Friday night!' He howled with laughter as he threw his little plane into a dangerously tight curve, doubling back towards the river. He stared at the radio indicator light and his addled brain realised that he was still transmitting.

'The crazy bastards have *done it*!' He shouted the last words at the top of his voice and thrashed the instrument fascia with his left fist. 'I woke up in a goddamned nightmare.'

He levelled out, having recovered the river, then followed it, travelling over the whole city, recognising none of it at first. 'Smoke's so bad . . .' His voice trailed off as he realised what he was seeing, a site that he had introduced to hundreds, maybe even thousands, of tourists over the last decade. 'Shevchenko's gone. The university's gone.'

The plane dropped 200 feet, barely off the ground now. Hammerick Bryan's subconscious mind knew what it was doing, even through an alcoholic haze. He stared in disbelief. Once, before the booze problem had returned to haunt him, he had been a man with a lover and even a real job. He knew some of his friends still worked there – *had* worked their. Dead. They must be dead. No one to listen to his avionics lectures now.

'Kids are dead, for God's sake. Maybe it was Chernobyl. Always said that damned thing would kill us all one day.' He flew on, recognising the shell of the great Philharmonic Concert Hall where Sophia had insisted on taking him for some 'culture'.

He could still remember her perfume. Not the name, but the scent: apples. She always smelt of apples, like the tree outside his window in spring, back home in Maryland. She was like his tree, strong inside, beautiful and so alive on the outside.

'No ordinary bomb could do this, no goddamned power station, either,' he told his radio. He felt lucid – maybe his liver had remembered how to process alcohol. But he doubted it – never known that particular organ to do any work. Or maybe it was just the shock that had cleared the haze.

He stared at the eleventh-century Golden Gate, one of the ancient entrances to the city. The land and buildings around it were blackened and shambolic, but somehow it had survived. The Gate stood as a lone reminder that life had once thrived in that great city, watched over by this gate for nearly nine centuries. Life would walk through those gates no more.

'Hanoi's back, guys,' Hammerick mumbled into the microphone, hoping his radio link would be heard in the only place he thought might be listening. 'I told ya it would come back.' His mind wandered, as he unconsciously put the plane into a climb to be sure he avoided a platoon of electricity pylons, standing to attention, their arms outstretched, at the city boundary. He thought of his buddies, the guys he had trained with: 44 of his fellow pilots had been taken; a lot of their gunners had bought it long before they had had a chance to experience Vietcong hospitality.

'Thirty-six thousand tons, we gave them. Twelve days of hell.' The children still talked to him, always had – even when he had learnt to talk back to them as his shrink had advised. They were there. Laughing and playing, then screaming and burning as his bombs took their homes, families and their lives. Always stood on a thick white line, waving their arms as if to frighten away a swarm of insects. The Twentieth Parallel, at Christmas, children burning. They walked through his mind when he least expected them. He could hear them now. Jeering, cajoling. He went through the moves, throttling back the great turbo-fan engines and giving the OK to the bomb bay: they were over their target. But there was no Hanoi to bomb today, no Kiev either.

He rubbed his dishevelled face, running his left hand over his brow and through the thick red mass of matted fibres that passed for hair. His right hand tightened its grip on the yoke, its knuckles aching as he tried to crush the plastic and steel control. 'Son of a bitch,' he said, using his usual condemnation of Nixon.

The plane's fuel light winked on. He hit the instrument panel again, but the light was stubborn. Climbing recklessly, he pushed the little old aircraft back to 2,000 feet and banked to the right, so he could circle over the black hole that had been a city.

'What the hell's going on?' he asked the radio, not bothering to toggle the hot key to allow an answer. He knew his dead friends would answer him soon enough. Whatever had happened, it was the end. He was sure of that.

Maybe there'd been an accident; maybe there were survivors. Maybe he could find Sophia. No point: she would never have him back; he was as good as dead to her anyway. He was an idiot, always had been. It was the damned kids – just wouldn't leave him alone. They were jealous of his life, his shrink had said, goading him, reminding him he had a life to live when they did not. Maybe this was his penance, being the last man alive in a world gone to hell.

Charlie 9194 flew on through the artificial cloud, grey and thick; snared in some impossible spider's web. He saw his airfield with its bizarre ex-Soviet concrete hangars, designed to protect their fighters from Western bombs, intact. He saw hangar 12, its door still open: the only place he could honestly have called home had kept him alive. A concrete cocoon, against a world on fire. Maybe he should have stayed there, safe on the ground. For what? It would have become his tomb sooner or later.

Dead. It was a dead place. No radio traffic, no noise at all, not even on the public channels. Not just the Kiev area. Europe was radio-silent – there was no long-wave signal, not a single very-high-frequency transmission either. Silence. Hammerick Bryan knew it was the end.

The fuel warning light blinked for forty minutes more. Then the children jeered no more.

Time after impact: +48 hours

The bell rang, its echoes heard on all sides of the green and blue ball. The great island masses, continents adrift in a sea of lava, water, grating over other land, folding, churning, turning this way and that. The great bell had rung its note of destruction and the world throbbed to its voice. Hours passed. The great slothful landmasses, ponderous and unaccustomed to change, reverberated, but nothing could be done to prevent the rifts. Red rivers of blood ran, bursting from their rocky veins, the world's body scarred and churned, its skin tortured, ruptured. Vast jets of golden rock melted to a liquid – great silica-rich rivers of blood 900 degrees Celsius, other arteries spewing viscous fluids hotter still at 1,100 degrees.

The world had shaken as a bell, its fabric shocked, the life it cared for dead and dying. Now it screamed out its rage and frustration as it too changed and fought to survive. Where cracks appeared it sealed them with its precious blood, bubbling and boiling over land, changing structure, building islands where none existed. Vast plumes of smoke poured into the already troubled biosphere. Black. Black as ink, dark as the cruellest heart, the skies changed as the great volcanic spew brought smoke and fumes. The breath of the under-world further changed the skies into a domain of acid and toxins.

Now the world would be new again. Cleansed by an aeon of fire and ice. Helios's rays sent back to their host, unwanted; the purification of ice would see off the unworthy. Just as the flames had purified, so the ice, too, would play its part. Slowly the islands formed, new folds and contours unfolded; all maps, all attempts of humankind to make order, began to fall away. The new age of ice beckoned.

Location: Refugee Camp Hermes, Australia.

Time after impact: +30 days

George Redwood's boot rested on the sign a little longer than was necessary. He carried his gun well. He had the posture of a man with training in the ways of death. Harry Saltzer and the lads followed him closely, watching the harsh shadows cast by the perimeter search-lights for any signs of movement. They moved with their impromptu leader almost as though he would protect them from anything. The outer perimeter fence had come down more easily than any of them had expected, but then it had been erected in a hurry. The media reckoned the government had had less than four weeks' notice to make preparations, in secret, before the impact.

The compounds were not popular with the locals, even less so now that rationing had been introduced – especially electricity: no one liked to be told when to use their power supplies. The government had faced a dilemma over the electricity question. Food was rationed, but restricting hours of electricity usage meant refrigeration was limited. With oil supplies severely limited due to the suspension of world trade after the loss of Bonn, Wall Street, London and other major international money markets, the government had to be sensible. Everyone knew that the refugees had brought their own supplies. Not happy to share them with the land of their benefactors, they kept them locked away in sovereign areas. The idiotic Australian government had given them the land to use for camps for their staff and soldiers, erected tents and cabins like a holiday resort. All of the jetsam of the old world – French, Brits, Italians, Spanish, all of the elite of course. No riffraff. That went down badly too. An awful lot of Australians could have called British and other European folk family. The person in the street was not best pleased to find that only aristocracy and bureaucracy had been seen as important enough to save.

Few would have supported such an initiative in Australia had it been publicly announced, and the national government knew it. Protests went on apace, inevitably bringing down the Labour government; no one was happy with the coalition, but at least it was better than having liars in office.

George was confident – the sort of confidence that comes with blind faith and bravado. The first shot took him in the shoulder. Before he had a chance to react the second tore his right calf muscle to shreds. The troops were on the invaders in seconds. Redwood's rash choice of Camp Hermes had been a mistake; the primary weapons cache of what survived of the British sovereign forces was well defended.

'Arms in the air,' a blackened face amongst a row of blackened faces ordered. It was unmistakably an East London accent.

'We didn't mean any harm. We just want some extra stuff for the kids, for God's sake. Some chocolate or something,' Harry told the mask.

'Plenty of food where you're going. Move!'

George Redwood's last thought before he passed out from pain was that no Australian would support the maintenance of such a force on home soil. The foreigners might have a place there, maybe settle and build homes, but they could not live like this. There had to be a change and soon. The foreigners had to accept that they were not going home. There was no home to go to.

Location: ice cave, Eisriesenwelt, Austrian Alps.
Pounding, struggling, her lungs would surely burst. Her feet were fleet, carrying her over the rocks, off the now oh-so-familiar tourist path, and into the dark areas, seldom trodden and little explored, save by ancient man, until the disaster.

She could not hear him any longer. He was still there, though, chasing her, determined to frighten her off, maybe kill her. She had no way of knowing. It was a terror she had had all her life, one that seldom left her. Her life had been replete with turning points. Places that she felt sure she could escape her past, shake off the horrors of childhood. But this! This damnable horror that had changed their lives, thrown their world into a living nightmare! No rescue parties, no saviours mounted aboard white chargers, no helicopters either, for that matter. They were utterly alone. She felt that no one on the whole Earth had survived, save for their enclave – she could hardly call it a community – they were all that there was. Or did she look on the dark side of matters a little too much? Whether humankind survived or not, she was determined to live out her life.

She saw it, the hole she had noticed the last time she and Jack had tried to find somewhere to be alone together – almost impossible with so many prying eyes. They had found it together and it had become their lovers' bolt hole. Few ventured this far from the light, but she was good at potholing, well experienced at the problems of seeing where only a cat might see, walking with purpose where others might grope.

She slipped as she tried to bring herself to a standstill, losing her footing on the ice. She grabbed a handhold, a piece of rock face that jutted into the cave. She dragged herself into the pothole – barely a metre high, but at least ten wide, it was a perfect hiding place: a dark hole.

She saw him. Not distinctly and clearly: just some barely tangible suggestion that he was there. He was foolish, old, at least 45, and cumbersome. His boots – she could hear them now that she had stopped running – pounded along the rock floor. Too fast.

208

She heard the tread of his boots as they gouged a trough in the ice that had taken her by surprise. She heard him fall, the sickening crack as his head hit rock, the just audible gurgling of air as it fled from her pursuer's lungs. He was dead and she was glad. Food was precious to all now, but sanctity was nonexistent. She had to cling to her dignity, to her privacy, if she was to stay sane. But many of the men, and indeed some of the women, seemed to view this scenario as one in which they could abandon all normal behaviour – just screw like animals with no care for those around them. They took what they wanted when they wanted it – food, clothes, self-esteem. They preyed on the innocent and the meek. Helen knew it was but a matter of time before something like this happened, but no one in her area of the caves had mentioned a death like this before.

She walked tall, knowing that her pursuer had found his own penance, and returned to her camp near the Ice Cathedral, as the tourist signs called it. There was little light. What there was came mostly from shoddy camp fires, fed on scraps and waste. It was unpleasant, but Iain and Jack believed that they could maintain a fire in the caves just as long as there were people there to produce the raw material. Jack had seen women in Indian shanties cooking on metal plates over dung fires. It worked. The smell faded into the background after a while, although, having been so far from the main camp for the first time in days, Helen noticed the smell immediately. She gagged, despite herself. She knew that the boys would be back soon.

Iain and Jack had been incredible: if it had not been for them she would surely have gone mad. Yet they too felt the same about her. She knew it, saw it in their eyes. She loved them both. If only she could show it.

They led two working parties. All of the men and most of the able-bodied women without children scoured the mountainside for food and resources. She had been a part of the first forays to the outside in the week after the impact. But the sights had been too terrible to bear. She was not weak, far from it. It was the grief – she could hardly bear it. She had proved to be a talented impromptu doctor – and that was needed more and more – yet several of the burns victims had already died. She had saved a good number of the children by spotting diarrhoea symptoms and feeding them as much sugar and salt water as they could spare. Some were dying, though – she knew it. If she thought it possible, she would swear that some were dying of a broken heart. Their womenfolk and children killed, many of the men were in a terrible state.

Iain had led a party to look for survivors, while Jack had taken charge of food gathering. For some reason the locals had taken to them, perhaps recognising that the two young men were good rock-hoppers, talented in the ways of the mountain. The irony was not lost on Helen: both were from the city, with a thirst for adventure. Both had become avid mountaineers – she felt sure they would have climbed to the moon if they could. She smiled, wondering how their fellow survivors would view the two British wildmen, as they called them now that their beards had grown, if they knew that they were city boys.

She walked past the cave mouth that led away from the light of the fire, back into the darkness where she knew her pursuer's body lay. He must have had rape on his mind, if he had a mind left after the madness of the last thirty days. She imagined his dead face contorted and shocked; it took her mind back to that first foray into the outside in week one.

Jack had been right about roots of trees and bushes surviving the firestorm on the mountain: amazingly, there was food if you dug hard enough. A couple of Austrian schoolboys had proved to be extremely talented mouse and vole catchers – they too had survived, buried underground in one of the few safe places left. Iain had learnt a lot from the rat-catchers, as Helen called them, and she was amazed to watch them work on that first trip. They used small coils of yarn or rope and formed little snares with slip knots that tightened around the legs and bodies of their unusual game. She had never quite adjusted to the idea of eating mice, but they offered protein, and hunger banished all sensibility.

That first expedition out of the caves had been horrifying, truly the stuff of nightmares. No one on Earth could see such sights and remain unchanged. They had walked for an hour, perhaps more, climbing over all of the rock and topsoil debris that had covered the mountain thanks to the aftershocks. Ever downward they walked. The Earth stank, reeked of burning material. It was as though someone had burnt the most appalling materials possible and dumped them on this mountain. It was like a great lion with its mane burnt off, blackened, bare and devoid of its regal finery.

At the end of that hour they had seen what she suspected Iain had expected to see. He was always quiet and brooding, but since the disaster he had become positively introverted whenever he had a problem to solve. He was quite capable of joviality, as he proved when he and Jack larked with the rat-catcher boys. Lord knew – there was little enough to be happy about. Still, Iain seemed to take on the entire weight of responsibility for the survival of these people. Maybe someone had to have that leadership edge, but she felt sure that no one else there, not one in the 1,000 people who had survived, seemed to view their predicament in quite that way. A few others had emerged as leaders and that was fine: many people were needed to find food and solve the problems they all faced. One, Marie Hollingsworth, another Englishwoman, had proved to be a brick. She was incredibly fit and walked for hours along the ridges in search of some sign of life across the Alps. She had found none, but never came back empty-handed, always managing to find some new source of food or combustible material. She had been the first to find unburnt, dry moss – great kindling for fires, and it had turned out to be easily the best 'toilet paper'. Marie had also taken charge of the latrines problem, earning her everyone's gratitude. She dug as much as, if not more than, the men who worked with her.

Helen had no doubt that Marie had saved their lives. Memories of school days filled her with horror as she remembered stories of nineteenth-century soldiers dying from dysentery because they had not realised the value of well-dug latrines. Horrible way to die. Marie was a real hero, not frightened to tackle the most difficult problems. It was all the more amazing that Marie did all of this despite the death of her husband. He had had some sort of attack, perhaps cardiac arrest – Helen had not been certain – on the second day. His death seemed to throw a switch in Marie's head, as though she were henceforth determined that no one else would die. Maybe that was her way of dealing with grief, or denying it. Whatever went on in Marie's mind, Helen knew that she had saved them all with her zeal for practical work and sound organisation.

That first expedition had proved to Helen that there was no God. At the end of the first hour the party reached the burnt-out shell of the last visitors' centre atop the mountain. A large one-level complex, rectilinear in ground plan, filled with restaurants that sold strange-looking hot dogs and outrageously overpriced photographic film – a place of happiness and relaxation – was a blackened, twisted abhorrence. The buildings smouldered still. Indeed, fire was a problem wherever they went, but it was not some roaring inferno, more a well-disguised hot spot. They were common. Many of the survivors had wandered off an established path looking for food, only to receive severe burns for their efforts; one young Frenchman, a tourist, died that way.

They walked to the cable-car shed. It was the second of two sets of cable cars that ascended the mountain from Tenneck. There they had stood, shoulder to shoulder, more than a hundred that day, looking out of the complex, down the mountain at the devastation that lay below. Until that moment, the murk, the almost tangible filth of the air, had been irrelevant to her. Now, however, she could *see* the air, or at least the smoke particulates it contained, shrouding everything in the world, it seemed, in a haze so dense that no colour, no direct light could penetrate it. The land, or at least that which could be discerned, was unrecognisable, ruined. Thousands of feet below them they could see the first-level cable-car complex, little more than a shambles. The second level had survived intact. Though scorched, the wires had held, but the cars were frozen in mid-journey.

Helen felt sure that she would never shake the horror of the mass that was discernible around the lower cable-car terminal. They all saw it, yet no one spoke of it. Many had obviously succeeded in taking a cable car to the first level only to be trapped there when the electrical power was disconnected by the fire. They died there, hundreds, maybe thousands, a mass grave, burnt, almost inhuman. Then there were the cable cars themselves. Two operated on the line: one travelled down the line while the other brought people up the mountain. The cars were stranded, suspended in space, thousands of feet above the still-smouldering landscape. The ascending cable car had a precious cargo – one that it was never able to deliver.

She would never forget them, their faces still distinguishable through the ash-covered windows. Sculptures frozen in a moment of terror. How long had they lived, suspended, trapped in space, with no hope of escape? They were not more than 500 metres from their destination, yet the cable's angle was not less than 25 degrees, far too steep to climb along. Some had tried, she felt sure: evident by the open hatch atop the car. Others must have lost their minds to the horror of it all. Trapped for ever, swaying in the terrible arctic winds that had descended. Gone mad for food, demented by the confinement. Helen could not shake their faces from her mind. She had seen much that might have driven a weaker person to distraction long before this disaster, yet those horrible deaths haunted her now.

212

Iain and Jack were due to return, hopefully with food. She shuddered again at the thought of the dead man, lost near her hiding place. He had become obsessional about Helen. Marie had noticed. Helen begged her not to confront the poor man, or to tell Iain and Jack, as their position as foreign 'guests' might become untenable were they to take rash action, especially now that the food seemed to be growing more scarce.

At least her troublesome shadow was gone. If only they could survive, maybe someone would come to help. Maybe.

Location: a hotel room, Canberra, Australian Capital Territory.

Time after impact: +150 days

It was cold. Not an ordinary cold, but something deeper – a cold from within, creeping across the soul. A cold of dread.

He opened his eyes. Dark angular silhouettes, moving in some unfathomable way, thrown into harsh contrast against a shocking white light, met his blurred gaze. Blink as he might, nothing would sharpen the image. A foul earthy stench, like swathes of rotten autumnal leaves, soured the air as he gulped in breath; breathing was now a barely remembered art. The sound of movement: damp rope on friction-smoothed wood. A weight stretching the rope to breaking point, making it spin this way and that, as if it were trying to break free of its load. Banks of a swirling ghostly ambience, hot breath on a dank day, obscured his vision, shrouding the dark shapes in a luminescent floating ether.

His breathing hoarse, he coughed, choking as his dry swollen tongue rasped against the roof of his mouth.

He coughed.

His diaphragm locked in spasm, his throat squeezing and releasing, desperate to push his enlarged tongue from his windpipe.

He coughed. His mind, struggling to free itself from the cold fingers of death, threw his limbs into spasm. His legs kicked against their restraints; his hands flailed at their bonds, desperate to scratch at his burning throat, tear a hole in his chest and free his screaming lungs.

He coughed. His head thrashed side to side, back and forth. His eyes lost their gift of sight – he could make nothing of the ghostly luminance and its strange silhouettes. Light faded to darkness, as his oxygen-starved brain began to shut down.

He could feel the guileless muscle in his chest pounding, pumping

213

useless carbon-dioxide-rich blood to his brain. The veins in his temples strained against their thin flesh covering, as if desperate to be free of their purpose. Blood vessels in his nasal passages ruptured, releasing the impossible levels of pressure in his veins. The strange grating rasps of the ropes above him ended their song, as his ear-drums burst, filling with blood as his foolish heart fought to feed his dying brain.

He coughed. His own tongue throttling his life as though a murderer were at his throat. He uttered a silent curse as all sense of life, all knowledge of his world around him, so recently discovered, so barely explored, ceased.

They cut the man down from his place of suicide, but they were too late to save him. He had been a man fraught with guilt, someone lost to history. His last days had been filled with anguish and sorrow. He had felt the blame of failure. Perhaps he could have done more, saved more. What did it matter? an adviser had asked him: the old ways were finished; it was a new world now.

It did matter.

It should matter to everyone, yet they were all so consumed with their own lives, surviving against the bitter cold and the impossible fogs, the unrelenting hostility of a world gone to ice and wind. Even the little communication that had been fostered with the United States had given scant hope. The most powerful government in the history of humankind, and all of its institutions, had fallen in a single day. America must recover – most of its people survived. But what of his own? They were no more, dust on the pages of a barely remembered history. He had died for them, for his guilt.

The death report showed his name and former occupation. His had been a title of pre-eminence, of power, but ultimately it mattered not. He was the leader of a dead land and so not a leader at all.

Location: somewhere in the Austrian Alps.

'What will it be like, Jack? I wonder what it'll be like.'

Jack smiled at Helen. His face was hard, rugged, noticeable even through his great mountain beard. His eyes were strong: they had seen much and he was the stronger for it now. He knew that some had cracked, gone under, and who could blame them? He was determined, however, that his son would be born. '*He'll* be fine and proud, just like his mum,' he said, smiling with his eyes. He knew that she took more strength from a smile than she could ever take from physical contact. He loved her with his eyes.

She smiled back, acknowledging his love, ignoring his mistake. She wondered about the lives they would lead, the world they had inherited.

'It'll be fine all right, Jack, but I'm banking on a girl,' Iain said, slapping Jack on the shoulder and putting a big arm round Helen.

'It's going to be damned hard now the ice and snows have come,' Jack said as he stared around the valley that lay before them. All around was a blanket of snow. Although the air was still hazy, they could distinguish darker areas that were obviously at lower levels, devoid of snow. 'I've a feeling that our little boy will know nothing but ice for a long time to come. But, whatever happens, we'll get through it. God knows we've done well enough already.'

They began their descent, hoping that the little rope and equipment that they had managed to retain would suffice on their climb down the great scorched mountain. They knew they had to leave. Food was hard to find now that the earth had frozen. They knew it would only get worse, as the mountain was not known, so the locals had said, to freeze this badly in winter, let alone summer. With a child on the way and resentment and sickness escalating within the protective womblike caves, they had to find another way.

Five months after their little training climb had begun, Helen, Jack and Iain the Troll began their descent. She looked at Iain, trying to find some trace of regret, but there was none. They, Helen and Jack, had expected him to stay in the caves with Susan. He loved her with all his heart, Helen was sure, but Susan was foolish, determined to stand her ground and remain in the caves with the little family she had left. Iain, ever practical, could not bring himself to stay behind and watch her die of hunger. He had given selflessly to the people of Tenneck – they knew it and loved him for it – but he wanted to live and find an answer to the question of what had happened to Earth and the people he had left at home.

Helen knew that her baby would be born – she could feel it in her bones. It was early days yet, less than two months, but she knew all would be well, even without the aid of modern medicine. Perhaps it was mothering instinct, or maybe this new life in an expanse of death had given her hope where there had been none. She did not know. What she knew for certain, however, was that the world her child would be born into would be as new to her as it would be to her baby. The land she knew, the world of her childhood, was no more.

Only time and the gods themselves knew what the future had in store.

AFTERWORD
By Jonathan Tate

For one astonishing week in July 1994 the world was treated to the greatest cosmic show in living memory. Over twenty fragments of the shattered Comet Shoemaker-Levy 9 collided with the planet Jupiter, one by one, plunging into the atmosphere of the giant planet at 134,000 miles per hour. The results were breathtaking. Massive plumes of super-hot gas were clearly visible from Earth, and immense dust clouds settled on the cloud tops of the wounded planet. Some of these dust scars were bigger than the Earth, and were clearly visible through even small telescopes. The blemishes took months to disperse, and a small band of scientists were finally able to say, 'See – comets do collide with planets!'

Impact Earth ably reminds us that we are very much a part of this cosmic shooting gallery. We are not safe, hidden on our little planet. Reading this book, I am reminded of the initial shock that I experienced during the Jupiter collision and my dawning realisation that Earth may soon face a terrible catastrophe.

One particular small telescope was trained on Jupiter each evening throughout that momentous week in 1994. My family and I were on a touring holiday in the northern United States – big-sky country – and conditions could hardly have been better. It is an odd phenomenon, but whenever you set up a telescope in a camp site you have an immediate crowd. Having seen the tiny image of Jupiter in the eyepiece, with the clear, dark spots that indicated the impact points, many people asked the obvious question: what would happen if such an event happened here on Earth?

Some months after the cataclysmic events over 477 million miles away, I decided to find out what measures were being taken by the international community to avoid such an eventuality here on Earth. The project seemed quite simple at the time, but the deeper I dug the less I found. It soon became apparent that the threat of cometary or asteroidal impact was being almost completely ignored by governments worldwide, despite the clear scientific evidence of its severity.

The Internet gave me access to the key scientists in the field, and they confirmed this impression. I met with, and talked to, many of them at an astronomical conference in Versailles in the summer of 1996 and, with their help, prepared a paper on the subject destined for the Secretary of State for Defence. The Ministry of Defence dismissed the problem as being beyond their capabilities to deal with, and consequently no action at all was planned. Approaches to the Department of Trade and Industry (who control the research councils) and the Department of the Environment fared no better.

The British National Space Centre, replying on behalf of the Secretary of State for Trade and Industry, assured me that 'the Government's view is that the scale of action taken so far is commensurate with the known level of the threat'. As the 'scale of action taken so far' is effectively zero, this statement was rather puzzling! Even more perplexing was the response from the Natural Environmental Research Council, which stated that NERC was concerned only with 'extreme natural events', and that consequently the impact threat was outside its remit!

So, on 1 January 1997, Spaceguard UK was launched with a number of stated aims:

- promoting and encouraging British activities involving the discovery and follow-up observations of near-Earth objects
- promoting the study of the physical and dynamic properties of asteroids and comets, with particular emphasis on near-Earth objects
- promoting the establishment of an international, ground-based surveillance network (the Spaceguard Project) for the discovery, observation and follow-up study of near-Earth objects
- providing a national United Kingdom information service to raise public awareness of the near-Earth-object threat, and technology available to predict and avoid dangerous impacts.

Within a year, all of the key figures in the field, both in the United Kingdom and abroad, could be counted among the members of Spaceguard UK. We are privileged to have Sir Crispin Tickell, Sir Arthur C Clarke, Professor Sir Bernard Lovell and Dr Patrick Moore as our patrons, and we maintain close links with the international

218

Spaceguard Foundation, based in Rome, and the other national organisations that are springing up worldwide. Thanks to our membership, Spaceguard UK is now the largest and most active organisation of its type in the world.

So, what has Spaceguard UK achieved so far? The events of 1998 have ensured that the level of public knowledge about the impact threat has been raised considerably, and Spaceguard UK has taken full advantage of all the opportunities offered to 'spread the word'. Members are often to be seen on television, heard on radio or read in popular and professional publications explaining the aims of Spaceguard UK in rational terms, as in *Impact Earth*. On the other hand, we have yet to achieve anything concrete. After years of banging our heads against the proverbial brick wall, the question remains: if the scientific evidence is so clear, why is so little being done about the gravest threat to our civilisation? It is clear that there are a number of reasons. First, the fact that there is a threat at all has come to the attention of the world in general only in the past decade or so. The evidence of cosmic impacts has been in plain view since the invention of the telescope, but the scale of planetary bombardment has become clear only since the advent of the space age. Past prejudice against catastrophist notions is disappearing in all fields of science as the evidence builds up, and the reality of major impacts is no longer in doubt.

Secondly, ancient humanity was quite convinced that cosmic influences had a significant part to play in its daily life and continued wellbeing. This conviction is clearly demonstrated in the stories and myths from around the world concerning conflict and disaster meted out from the skies. This catastrophist view of the cosmos reigned supreme until the Age of Reason, when Newtonian principles turned the unknown and unpredictable universe into a benign, mechanical system and Darwinism spawned the concept of gradual evolution over extended periods of time. In the resulting predictable, gradualist cosmos there was no place for catastrophism or major, sudden changes in the global environment. Since the late 1980s the realisation that Darwinian evolution has almost certainly been punctuated by massive catastrophic events, causing major redirection in biological and geographic evolution, has led to the breaking of many scientific paradigms. The move from gradualism towards catastrophism is causing a major rethink of many cherished ideas.

Thirdly, some still maintain that nothing can be done about the

219

problem, so why bother? Until the dawn of the space age humankind was truly helpless in the face of cosmic bombardment, so there was little to be gained from worrying about the problem. However, the advances in technology that have occurred during the past two decades or so have changed the situation radically. Surveillance technology now allows the detection and tracking of threatening objects, while spacecraft technology has advanced sufficiently to allow the interception of such bodies. Measures to deflect threatening bodies have been developed, involving the use of nuclear weapons, but emerging technologies may provide more 'environmentally friendly' alternatives. The problem can be dealt with now, albeit in a fairly crude fashion.

Of course, money is always a stumbling block. NEO surveillance, tracking and mitigation programmes will cost money. While the first two are relatively cheap, any expenditure needs detailed and credible justification in these days of fiscal stringency. The threat to humankind from asteroidal or cometary impact is not high enough on anybody's priority list for the required funding. In part, this is due to a lack of knowledge relating to the threat, and also to the perception that other contingencies are more important.

The blame for inaction cannot be placed exclusively at the door of the politicians. Within the scientific community there is still disagreement, some acrimonious, over the nature and extent of the threat. It is perfectly natural for scientists to disagree – indeed, that is the nature of the scientific method. However, few would dispute at least the possibility of a significant threat, and, given the possible consequences, it is not justifiable to oppose programmes to assess that possibility. That would be playing dice with the survival of the human species. Many scientific bodies oppose research into the NEO threat on the grounds that such programmes might divert funding from their particular fields. While this is perfectly understandable from a narrow perspective, it is an abrogation of the responsibility of science to safeguard humankind, or at least to alert it to threats to its wellbeing.

However, the most intractable problem faced by those trying to get something done involves the question of responsibility: exactly who should be responsible for planetary defence? The subject is a multidisciplinary undertaking – astronomers have been at the forefront of the search for and detection of NEOs, but planetary scientists, geologists, palaeontologists, biologists, physicists and

220

many others have been deeply involved in piecing together the jigsaw that has resulted in our current state of knowledge.

But is the problem strictly scientific? Scientists are concerned with the acquisition and interpretation of new data. To study asteroids and comets the researcher needs to study only a representative sample; there is no need to find them all. But a planetary defence programme would have to do just that. The funding and resources required to detect and track all NEOs cannot therefore be justified on scientific research grounds. Defence is usually the prerogative of the military, but there is some resistance from the defence establishment to becoming involved in planetary defence. The excuses hide the real reason – money. Defence budgets are stretched to the limit without having a new drain on already scarce resources. So, neither the scientific nor military community is willing to take responsibility for planetary defence. Governments will have to come to some decision sooner or later, preferably before the event.

Hopefully, by the time you read this, those responsible for the welfare of their peoples will have taken some sensible and prudent decisions. The major objective of Spaceguard UK is to disband itself, as soon as a sensible surveillance system is in place and operating. I look forward to that day, when we will know that our home is safe, with hope and optimism.

If *Impact Earth* has stirred these same concerns within you, I urge you to write to your Member of Parliament and contact your national Spaceguard pressure group, but, above all, add your own voice to the growing chorus that now demands action. We can make a difference.

Major Jonathan Tate
Spaceguard UK
25 December 1998

If you would like to know more about the impact threat, or about Spaceguard UK, please contact:
Spaceguard UK
Cygnus Lodge
High Street
Figheldean
Wiltshire SP4 8JT
Telephone: 01980 671380; Fax: 01980 671381; Email: Spaceguard-@dial.pipex.com; Website: http://ds.dial.pipex.com/spaceguard/

Appendix

*Predicted comet and asteroid near-Earth approaches, including plan-
etoids that represent potential impact hazards.*

FORTHCOMING CLOSE APPROACHES TO THE EARTH
The following table lists the predicted minor-planet and comet
encounters to within 0.2 AU of the Earth during the next 33 years
(from the month of June 1999). Objects with very uncertain orbits
are excluded from this listing, as are recently discovered objects
whose orbits have been computed without consideration of planetary
perturbations. If the gravitational pull of the gas giants (Jupiter and
Saturn) or other planetoidal gravitational effects were to perturb the
orbits of the listed bodies, we might be lulled into a false sense
of security. Consequently, recent discoveries need more follow-up
observation before being properly catalogued. That said, the current
listings and the distances quoted for listed objects are far from
absolute, indeed the orbits of objects recorded during only one oppo-
sitional pass (such as 1991 DB and 1986 JK) must be seen as nominal
only and prone to revision. In short, although unlikely, it is possible
that objects that are predicted to avoid Earth may actually impact.
As ever, more follow-up observation is needed. The 'heroic' efforts
of the world's NEO hunters would be undermined without the continued efforts of the Minor Planet Center. As observation increases,
so the MPC's role will continue to grow in significance and it is
hoped that increases in funding will follow suit.

Object and name		Date of encounter	Projected vicinity to Earth (AU)	No. observations to trace orbital arc (revealing the path taken by the object as it travels around the solar system)	Reference
(6489)	Golevka	1999 June	0.0500	2 oppositions, 1991–1995	MPC 25418
	1989 VA	1999 Nov.	0.1938	5 oppositions, 1989–1997	MPC 31005
	1991 DB	2000 Mar.	0.1017	1-opposition, arc = 147 days	MPC 20820
	1986 JK	2000 July	0.1143	1-opposition, arc = 179 days	MPC 24117
	1991 BB	2000 July	0.1662	4 oppositions, 1991–1996	MPC 28316
(4486)	Mithra	2000 Aug.	0.0465	5 oppositions, 1974–1996	MPC 28565
(2100)	Ra-Shalom	2000 Sept.	0.1896	9 oppositions, 1975–1997	MPC 30432
(2340)	Hathor	2000 Oct.	0.1970	5 oppositions, 1976–1997	MPC 30890
(4179)	Toutatis	2000 Oct.	0.0739	8 oppositions, 1934–1996	MPC 28565
(4183)	Cuno	2000 Dec.	0.1426	9 oppositions, 1986–1998	MPC 32296
(4688)	1980 WF	2001 Jan.	0.1701	2 oppositions, 1980–1991	MPC 17609
(4034)	1986 PA	2001 Apr.	0.1465	4 oppositions, 1986–1997	MPC 30250
(3103)	Eger	2001 Aug.	0.1161	8 oppositions, 1982–1997	MPC 32296
	1987 QB	2001 Aug.	0.1629	1-opposition, arc = 171 days	MPC 14023
(3362)	Khufu	2001 Dec.	0.1597	10 oppositions, 1984–1996	MPC 28056
(7341)	1991 VK	2002 Jan.	0.0718	4 oppositions, 1991–1996	MPC 28573
(4660)	Nereus	2002 Jan.	0.0290	4 oppositions, 1981–1993	MPC 23852
(3361)	Orpheus	2002 Jan.	0.1695	7 oppositions, 1982–1998	MPC 32296
(5604)	1992 FE	2002 June	0.0768	4 oppositions, 1976–1993	MPC 22476
(2101)	Adonis	2002 June	0.1610	4 oppositions, 1936–1995	MPC 24887
	1989 VA	2002 Oct.	0.1771	5 oppositions, 1989–1997	MPC 31005

(3362)	Khufu	2002 Dec.	0.1498	10 oppositions, 1984–1996	MPC 28056
(5381)	Sekhmet	2003 May	0.1285	4 oppositions, 1991–1995	MPC 25192
(6489)	Golevka	2003 May	0.0923	2 oppositions, 1991–1995	MPC 25418
(2100)	Ra-Shalom	2003 Aug.	0.1745	9 oppositions, 1975–1997	MPC 30432
(7336)	1989 RS1	2003 Sept.	0.1903	3 oppositions, 1982–1996	MPC 28571
(2063)	Bacchus	2003 Sept.	0.1217	7 oppositions, 1977–1996	MPC 26885
(4197)	1982 TA	2003 Oct.	0.1866	8 oppositions, 1954–1996	MPC 28057
	1989 UQ	2003 Oct.	0.1496	4 oppositions, 1954–1996	MPC 28613
(3362)	Khufu	2003 Dec.	0.1946	10 oppositions, 1984–1996	MPC 28056
(6239)	Minos	2004 Feb.	0.0564	4 oppositions, 1983–1994	MPC 27892
(4179)	Toutatis	2004 Sept.	0.0104	8 oppositions, 1934–1996	MPC 28565
(3908)	1980 PA	2004 Nov.	0.1394	3 oppositions, 1980–1997	MPC 29582
(7753)	1988 XB	2004 Nov.	0.0729	5 oppositions, 1988–1997	MPC 30258
(7350)	1993 VA	2005 Feb.	0.1427	4 oppositions, 1963–1996	MPC 28575
(6611)	1993 VW	2005 Apr.	0.0862	4 oppositions, 1982–1995	MPC 25726
(4544)	Xanthus	2005 May	0.1938	2 oppositions, 1989–1990	MPC 16572
(5660)	1974 MA	2005 Aug.	0.1874	5 oppositions, 1974–1996	MPC 27291
	1977 VA	2005 Oct.	0.1362	1–opposition, arc = 93 days	MPC 22073
(1862)	Apollo	2005 Nov.	0.0752	12 oppositions, 1930–1998	MPC 32295
(3361)	Orpheus	2006 Jan.	0.1597	7 oppositions, 1982–1998	MPC 32296
(5797)	Bivoj	2006 Feb.	0.1899	2 oppositions, 1980–1993	MPC 22936
(3103)	Eger	2006 Aug.	0.1284	8 oppositions, 1982–1997	MPC 32296
(4450)	Pan	2006 Sept.	0.1464	5 oppositions, 1987–1994	MPC 23960
(7341)	1991 VK	2007 Jan.	0.0679	4 oppositions, 1991–1996	MPC 28573
(5011)	Ptah	2007 Jan.	0.1982	6 oppositions, 1960–1997	MPC 30250

Object and name	Date of encounter	Projected vicinity to Earth (AU)	No. observations to trace orbital arc (revealing the path taken by the object as it travels around the solar system)	Reference
(1862) Apollo	2007 May	0.0714	12 oppositions, 1930–1998	MPC 32295
(2340) Hathor	2007 Oct.	0.0600	5 oppositions, 1976–1997	MPC 30890
1989 VA	2007 Oct.	0.1552	5 oppositions, 1989–1997	MPC 31005
1989 UR	2007 Nov.	0.0710	3 oppositions, 1989–1998	MPC 32741
(3200) Phaethon	2007 Dec.	0.1209	10 oppositions, 1983–1996	MPC 28056
(1685) Toro	2008 Jan.	0.1964	21 oppositions, 1948–1997	MPC 32295
(4450) Pan	2008 Feb.	0.0408	5 oppositions, 1987–1994	MPC 23960
(6037) 1988 EG	2008 Mar.	0.1667	5 oppositions, 1988–1998	MPC 31414
(1620) Geographos	2008 Mar.	0.1251	19 oppositions, 1951–1994	MPC 25514
1991 DG	2008 Aug.	0.1753	3 oppositions, 1991–1996	MPC 28085
(8567) 1996 HW1	2008 Sept.	0.1351	4 oppositions, 1980–1998	MPC 31512
(4179) Toutatis	2008 Nov.	0.0502	8 oppositions, 1934–1996	MPC 28565
1991 DB	2009 Mar.	0.1168	1–opposition, arc = 147 days	MPC 20820
1991 JW	2009 May	0.0812	3 oppositions, 1991–1997	MPC 30091
(9162) 1987 OA	2009 Aug.	0.1060	4 oppositions, 1987–1998	MPC 32415
(8566) 1996 EN	2009 Aug.	0.1942	3 oppositions, 1996–1998	MPC 31511
(3361) Orpheus	2009 Dec.	0.1384	7 oppositions, 1982–1998	MPC 32296
(4486) Mithra	2010 Mar.	0.1889	5 oppositions, 1974–1996	MPC 28565
(6239) Minos	2010 Aug.	0.0985	4 oppositions, 1983–1994	MPC 27892
1989 UQ	2010 Oct.	0.1533	4 oppositions, 1954–1996	MPC 28613
(3838) Epona	2010 Nov.	0.1973	5 oppositions, 1986–1996	MPC 28565

	1991 JW	2010 Nov.	0.0952	3 oppositions, 1991–1997	MPC 30091
	1990 SS	2011 Mar.	0.0994	1–opposition, arc = 205 days	MPC 21974
(3988)	1986 LA	2011 June	0.1853	4 oppositions, 1986–1997	MPC 29074
(3103)	Eger	2011 Aug.	0.1528	8 oppositions, 1982–1997	MPC 32296
	1988 TA	2011 Aug.	0.1972	1–opposition, arc = 63 days	MPC 29941
(7341)	1991 VK	2012 Jan.	0.0650	4 oppositions, 1991–1996	MPC 28573
(433)	Eros	2012 Jan.	0.1787	42 oppositions, 1893–1993	MPC 24086
(4183)	Cuno	2012 May	0.1218	9 oppositions, 1986–1998	MPC 32296
(4581)	Asclepius	2012 Aug.	0.1079	2 oppositions, 1989–1990	MPC 16864
(4769)	Castalia	2012 Aug.	0.1135	4 oppositions, 1989–1995	MPC 25518
	1989 VA	2012 Nov.	0.1644	5 oppositions, 1989–1997	MPC 31005
(4179)	Toutatis	2012 Dec.	0.0463	8 oppositions, 1934–1996	MPC 28565
(3752)	Camillo	2013 Feb.	0.1478	5 oppositions, 1985–1995	MPC 25518
(7888)	1993 UC	2013 Mar.	0.1260	5 oppositions, 1989–1997	MPC 30641
(4034)	1986 PA	2013 Apr.	0.1528	4 oppositions, 1986–1997	MPC 30250
	1988 TA	2013 May	0.0411	1–opposition, arc = 63 days	MPC 29941
(7753)	1988 XB	2013 July	0.1183	5 oppositions, 1988–1997	MPC 30258
(6037)	1988 EG	2013 Aug.	0.1847	5 oppositions, 1988–1998	MPC 31414
(4581)	Asclepius	2013 Aug.	0.1188	2 oppositions, 1989–1990	MPC 16864
(6063)	Jason	2013 Nov.	0.0790	4 oppositions, 1984–1995	MPC 25518
(3361)	Orpheus	2013 Dec.	0.1032	7 oppositions, 1982–1998	MPC 32296
(2062)	Aten	2014 Jan.	0.1463	6 oppositions, 1955–1995	MPC 24719
(2340)	Hathor	2014 Oct.	0.0482	5 oppositions, 1976–1997	MPC 30890
	1987 WC	2014 Nov.	0.1249	2 oppositions, 1987–1995	MPC 25528
(2062)	Aten	2015 Jan.	0.1864	6 oppositions, 1955–1995	MPC 24719

Object and name		Date of encounter	Projected vicinity to Earth (AU)	No. observations to trace orbital arc (revealing the path taken by the object as it travels around the solar system)	Reference
(2063)	Bacchus	2015 Apr.	0.1952	7 oppositions, 1977–1996	MPC 26885
(5381)	Sekhmet	2015 May	0.1613	4 oppositions, 1991–1995	MPC 25192
(1566)	Icarus	2015 June	0.0545	14 oppositions, 1949–1982	MPC 8665
(7822)	1991 CS	2015 Sept.	0.1596	2 oppositions, 1991–1997	MPC 30437
(1685)	Toro	2016 Jan.	0.1566	21 oppositions, 1948–1997	MPC 32295
(7822)	1991 CS	2016 Feb.	0.1683	2 oppositions, 1991–1997	MPC 30437
(7350)	1993 VA	2016 Mar.	0.1533	4 oppositions, 1963–1996	MPC 28575
(3103)	Eger	2016 Aug.	0.1894	8 oppositions, 1982–1997	MPC 32296
(2100)	Ra-Shalom	2016 Oct.	0.1499	9 oppositions, 1975–1997	MPC 30432
(5143)	Heracles	2016 Nov.	0.1470	7 oppositions, 1962–1996	MPC 28057
(2102)	Tantalus	2016 Dec.	0.1375	5 oppositions, 1975–1994	MPC 24371
(7341)	1991 VK	2017 Jan.	0.0647	4 oppositions, 1991–1996	MPC 28573
(5604)	1992 FE	2017 Feb.	0.0336	4 oppositions, 1976–1993	MPC 22476
(6063)	Jason	2017 May	0.0985	4 oppositions, 1984–1995	MPC 25518
(3122)	Florence	2017 Sept.	0.0472	8 oppositions, 1981–1995	MPC 24720
(5189)	1990 UQ	2017 Sept.	0.0610	4 oppositions, 1990–1996	MPC 26727
(2329)	Orthos	2017 Sept.	0.1582	10 oppositions, 1976–1998	MPC 32295
	1989 UQ	2017 Oct.	0.1563	4 oppositions, 1954–1996	MPC 28613
	1989 UP	2017 Nov.	0.0549	2 oppositions, 1989–1995	MPC 24894
	1989 VA	2017 Nov.	0.1779	5 oppositions, 1989–1997	MPC 31005
(3361)	Orpheus	2017 Nov.	0.0607	7 oppositions, 1982–1998	MPC 32296

Number	Name	Date		Oppositions	MPC
(3200)	Phaethon	2017	Dec.	0.0690	10 oppositions, 1983–1996	MPC 28056
(3752)	Camillo	2018	Feb.	0.1378	5 oppositions, 1985–1995	MPC 25518
	1991 DB	2018	Mar.	0.1749	1-opposition, arc = 147 days	MPC 20820
(1981)	Midas	2018	Mar.	0.0896	7 oppositions, 1973–1992	MPC 22575
(4953)	1990 MU	2018	Nov.	0.1068	9 oppositions, 1974–1994	MPC 23770
(2340)	Hathor	2019	Jan.	0.1441	5 oppositions, 1976–1997	MPC 30890
(4581)	Asclepius	2019	Mar.	0.1381	2 oppositions, 1989–1990	MPC 16864
(5604)	1992 FE	2019	June	0.1318	4 oppositions, 1976–1993	MPC 22476
	1989 UR	2019	June	0.1870	3 oppositions, 1989–1998	MPC 32741
(1620)	Geographos	2019	Aug.	0.1373	19 oppositions, 1951–1994	MPC 25514
(2100)	Ra-Shalom	2019	Sept.	0.1797	9 oppositions, 1975–1997	MPC 30432
(4581)	Asclepius	2020	Mar.	0.0705	2 oppositions, 1989–1990	MPC 16864
	1991 DG	2020	Apr.	0.0845	3 oppositions, 1991–1996	MPC 28085
(8014)	1990 MF	2020	July	0.0547	2 oppositions, 1990–1997	MPC 30855
(9162)	1987 OA	2020	Aug.	0.1823	4 oppositions, 1987–1998	MPC 32415
(5645)	1990 SP	2020	Oct.	0.1420	3 oppositions, 1990–1993	MPC 22479
(7753)	1988 XB	2020	Nov.	0.0662	5 oppositions, 1988–1997	MPC 30258
(5879)	1992 CH1	2021	Feb.	0.1987	4 oppositions, 1992–1997	MPC 30684
(5189)	1990 UQ	2021	May	0.0682	4 oppositions, 1990–1996	MPC 26727
(7822)	1991 CS	2021	Aug.	0.1454	2 oppositions, 1991–1997	MPC 30437
(2340)	Hathor	2021	Nov.	0.1606	5 oppositions, 1976–1997	MPC 30890
(3361)	Orpheus	2021	Nov.	0.0386	7 oppositions, 1982–1998	MPC 32296
(4660)	Nereus	2021	Dec.	0.0263	4 oppositions, 1981–1993	MPC 23852
(7482)	1994 PC1	2022	Jan.	0.0132	4 oppositions, 1974–1997	MPC 29085
(7341)	1991 VK	2022	Jan.	0.0640	4 oppositions, 1991–1996	MPC 28573

Object and name		Date of encounter	Projected vicinity to Earth (AU)	No. observations to trace orbital arc (revealing the path taken by the object as it travels around the solar system)	Reference
(7335)	1989 JA	2022 May	0.0269	3 oppositions, 1989–1996	MPC 28570
(5693)	1993 EA	2022 May	0.0686	5 oppositions, 1984–1994	MPC 23116
	1989 ML	2022 July	0.0995	3 oppositions, 1989–1996	MPC 27324
	1991 BB	2022 July	0.1293	4 oppositions, 1991–1996	MPC 28316
(2100)	Ra-Shalom	2022 Aug.	0.1873	9 oppositions, 1975–1997	MPC 30432
	1989 VA	2022 Nov.	0.1901	5 oppositions, 1989–1997	MPC 31005
	1990 HA	2022 Nov.	0.0909	2 oppositions, 1989–1990	MPC 31784
(4486)	Mithra	2023 Apr.	0.1627	5 oppositions, 1974–1996	MPC 28565
(4769)	Castalia	2023 Aug.	0.1100	4 oppositions, 1989–1995	MPC 25518
(6037)	1988 EG	2023 Aug.	0.0407	5 oppositions, 1988–1998	MPC 31414
(4544)	Xanthus	2023 Oct.	0.1956	2 oppositions, 1989–1990	MPC 16572
(7350)	1993 VA	2023 Nov.	0.1294	4 oppositions, 1963–1996	MPC 28575
	1989 DA	2024 Jan.	0.1354	1-opposition, arc = 89 days	MPC 14795
(1685)	Toro	2024 Jan.	0.1331	21 oppositions, 1948–1997	MPC 32295
(2063)	Bacchus	2024 Mar.	0.1200	7 oppositions, 1977–1996	MPC 26885
	1986 JK	2024 May	0.1166	1-opposition, arc = 179 days	MPC 24117
	1989 UQ	2024 Oct.	0.1559	4 oppositions, 1954–1996	MPC 28613
(887)	Alinda	2025 Jan.	0.0822	21 oppositions, 1918–1998	MPC 32295
(6239)	Minos	2025 Feb.	0.0991	4 oppositions, 1983–1994	MPC 27892
(4034)	1986 PA	2025 Apr.	0.1568	4 oppositions, 1986–1997	MPC 30250
(9058)	1992 JB	2025 Apr.	0.1022	4 oppositions, 1992–1998	MPC 32187

(2100)	Ra-Shalom	2025 Aug.	0.1729	9 oppositions, 1975–1997	MPC 30432
	1989 VA	2025 Oct.	0.1846	5 oppositions, 1989–1997	MPC 31005
(3361)	Orpheus	2025 Nov.	0.0379	7 oppositions, 1982–1998	MPC 32296
(2340)	Hathor	2026 Jan.	0.1044	5 oppositions, 1976–1997	MPC 30890
(1943)	Anteros	2026 May	0.1322	11 oppositions, 1973–1997	MPC 32295
(1620)	Geographos	2026 Aug.	0.1704	19 oppositions, 1951–1994	MPC 25514
	1989 AZ	2026 Sept.	0.1898	1-opposition, arc = 39 days	MPC 15069
	1990 HA	2026 Dec.	0.0926	2 oppositions, 1989–1990	MPC 31784
(5693)	1993 EA	2026 Dec.	0.0531	5 oppositions, 1984–1994	MPC 23116
(7341)	1991 VK	2027 Jan.	0.0681	4 oppositions, 1991–1996	MPC 28573
(4769)	Castalia	2027 Apr.	0.1288	4 oppositions, 1989–1995	MPC 25518
	1991 JW	2027 May	0.1000	3 oppositions, 1991–1997	MPC 30091
(4953)	1990 MU	2027 June	0.0309	9 oppositions, 1974–1994	MPC 23770
(5011)	Ptah	2027 Nov.	0.1911	6 oppositions, 1960–1997	MPC 30250
	1989 UR	2028 June	0.0888	3 oppositions, 1989–1998	MPC 32741
	1991 EE	2028 Sept.	0.0968	4 oppositions, 1991–1998	MPC 32743
	1985 WA	2028 Oct.	0.1209	1-opposition, arc = 97 days	MPC 21970
	1991 JW	2028 Nov.	0.1345	3 oppositions, 1991–1997	MPC 30091
(7753)	1988 XB	2029 July	0.1148	5 oppositions, 1988–1997	MPC 30258
(3361)	Orpheus	2029 Nov.	0.0842	7 oppositions, 1982–1998	MPC 32296
(5731)	Zeus	2029 Dec.	0.0987	5 oppositions, 1988–1993	MPC 22675
(5011)	Ptah	2030 Feb.	0.1741	6 oppositions, 1960–1997	MPC 30250
(1917)	Cuyo	2030 Oct.	0.0897	13 oppositions, 1954–1998	MPC 32295
	1989 VA	2030 Oct.	0.1549	5 oppositions, 1989–1997	MPC 31005
	1990 HA	2030 Dec.	0.1774	2 oppositions, 1989–1990	MPC 31784

Object and name	Date of encounter	Projected vicinity to Earth (AU)	No. observations to trace orbital arc (revealing the path taken by the object as it travels around the solar system)	Reference
(4660) Nereus	2031 Mar.	0.1150	4 oppositions, 1981–1993	MPC 23852
1990 SS	2031 Mar.	0.1422	1–opposition, arc = 205 days	MPC 21974
1987 SF3	2031 July	0.1687	1–opposition, arc = 51 days	MPC 23513
(6239) Minos	2031 Aug.	0.1294	4 oppositions, 1983–1994	MPC 27892
(2063) Bacchus	2031 Sept.	0.1263	7 oppositions, 1977–1996	MPC 26885
1989 UQ	2031 Oct.	0.1588	4 oppositions, 1954–1996	MPC 28613
(5645) 1990 SP	2031 Nov.	0.0985	3 oppositions, 1990–1993	MPC 22479

List is based on data produced by the Minor Planet Center, Harvard University, Cambridge, Mass., USA.

PHA CLOSE APPROACHES TO THE EARTH

The following table lists the predicted encounters by Potentially Hazardous Asteroids (PHAs) to within 0.05 AU of the Earth from the start of 1998 through to the end of the 21st century. Objects with very uncertain orbits are excluded from this listing, as are recently discovered objects whose orbits have again been computed without consideration of planetary perturbations. The distances quoted are from the nominal orbit solutions in the cited references and can be quite uncertain, particularly for one-opposition objects.

Object and name		Date of encounter	Projected vicinity to Earth (AU)	No. of observations to trace orbital arc (revealing the path taken by the object as it travels around the solar system)	Reference
(2340)	Hathor	2086 Oct.	0.0059	5 oppositions, 1976–1997	MPC 30890
(2340)	Hathor	2069 Oct.	0.0066	5 oppositions, 1976–1997	MPC 30890
(4660)	Nereus	2060 Feb.	0.0080	4 oppositions, 1981–1993	MPC 23852
	1988 TA	2053 Oct.	0.0088	1–opposition, arc = 63 days	MPC 29941
(4179)	Toutatis	2004 Sept.	0.0104	8 oppositions, 1934–1996	MPC 28565
(4581)	Asclepius	2051 Mar.	0.0122	2 oppositions, 1989–1990	MPC 16864
(7482)	1994 PC1	2022 Jan.	0.0132	4 oppositions, 1974–1997	MPC 29085
(4660)	Nereus	2071 Feb.	0.0149	4 oppositions, 1981–1993	MPC 23852
	1989 UQ	2093 Aug.	0.0160	4 oppositions, 1954–1996	MPC 28613
(3200)	Phaethon	2093 Dec.	0.0194	10 oppositions, 1983–1996	MPC 28056
(4179)	Toutatis	2069 Nov.	0.0199	8 oppositions, 1934–1996	MPC 28565
(3361)	Orpheus	2091 Apr.	0.0211	7 oppositions, 1982–1998	MPC 32296
(3362)	Khufu	2045 Aug.	0.0212	10 oppositions, 1984–1996	MPC 28056
(4953)	1990 MU	2058 June.	0.0229	9 oppositions, 1974–1994	MPC 23770

Object and name	Date of encounter	Projected vicinity to Earth (AU)	No. observations to trace orbital arc (revealing the path taken by the object as it travels around the solar system)	Reference
1989 UQ	2087 Dec.	0.0240	4 oppositions, 1954–1996	MPC 28613
(2340) Hathor	2045 Oct.	0.0242	5 oppositions, 1976–1997	MPC 30890
(6037) 1988 EG	2041 Feb.	0.0244	5 oppositions, 1988–1998	MPC 31414
(7822) 1991 CS	2065 Aug.	0.0250	2 oppositions, 1991–1997	MPC 30437
(4769) Castalia	2046 Aug.	0.0251	4 oppositions, 1989–1995	MPC 25518
(8566) 1996 EN	2070 Sept.	0.0251	3 oppositions, 1996–1998	MPC 31511
(4660) Nereus	2021 Dec.	0.0263	4 oppositions, 1981–1993	MPC 23852
1989 UP	2050 Dec.	0.0267	2 oppositions, 1989–1995	MPC 24894
(7335) 1989 JA	2022 May	0.0269	3 oppositions, 1989–1996	MPC 28570
(3671) Dionysus	2085 June	0.0280	4 oppositions, 1984–1997	MPC 29889
(4660) Nereus	2002 Jan.	0.0290	4 oppositions, 1981–1993	MPC 23852
(8566) 1996 EN	2033 Sept.	0.0303	3 oppositions, 1996–1998	MPC 31511
1991 EE	2065 Sept.	0.0305	4 oppositions, 1991–1998	MPC 32743
(4953) 1990 MU	2027 June.	0.0309	9 oppositions, 1974–1994	MPC 23770
(6037) 1988 EG	1998 Feb.	0.0318	5 oppositions, 1988–1998	MPC 31414
1989 UP	2078 Dec.	0.0326	2 oppositions, 1989–1995	MPC 24894
(1862) Apollo	2082 May.	0.0334	12 oppositions, 1930–1998	MPC 32295
(5604) 1992 FE	2017 Feb.	0.0336	4 oppositions, 1976–1993	MPC 22476
(6037) 1988 EG	2079 Aug.	0.0342	5 oppositions, 1988–1998	MPC 31414
1989 UR	2053 Nov.	0.0351	3 oppositions, 1989–1998	MPC 32741
(1862) Apollo	2046 Nov.	0.0353	12 oppositions, 1930–1998	MPC 32295

(1566)	Icarus	2090 June.	0.0357	14 oppositions, 1949–1982	MPC 8665
(2101)	Adonis	2036 Feb.	0.0357	4 oppositions, 1936–1995	MPC 24887
(4450)	Pan	2060 Feb.	0.0369	5 oppositions, 1987–1994	MPC 23960
(3361)	Orpheus	2025 Nov.	0.0379	7 oppositions, 1982–1998	MPC 32296
(3757)	1982 XB	2074 Dec.	0.0379	3 oppositions, 1982–1997	MPC 30890
(3361)	Orpheus	2021 Nov.	0.0386	7 oppositions, 1982–1998	MPC 32296
(7822)	1991 CS	2090 Sept.	0.0391	2 oppositions, 1991–1997	MPC 30437
(8014)	1990 MF	2080 Sept.	0.0393	2 oppositions, 1990–1997	MPC 30855
(6037)	1988 EG	2023 Aug.	0.0407	5 oppositions, 1988–1998	MPC 31414
(4450)	Pan	2008 Feb.	0.0408	5 oppositions, 1987–1994	MPC 23960
	1988 TA	2013 May	0.0411	1-opposition, arc = 63 days	MPC 29941
(2102)	Tantalus	2038 Dec.	0.0443	5 oppositions, 1975–1994	MPC 24371
(4179)	Toutatis	2012 Dec.	0.0463	8 oppositions, 1934–1996	MPC 28565
(4486)	Mithra	2000 Aug.	0.0465	5 oppositions, 1974–1996	MPC 28565
(1620)	Geographos	2051 Aug.	0.0479	19 oppositions, 1951–1994	MPC 25514
(2340)	Hathor	2014 Oct.	0.0482	5 oppositions, 1976–1997	MPC 30890
(1862)	Apollo	2048 May.	0.0498	12 oppositions, 1930–1998	MPC 32295
(6489)	Golevka	1999 June	0.0500	2 oppositions, 1991–1995	MPC 25418

Glossary

ACCRETION
Stars, planets and small bodies such as asteroids are thought to form by 'accretion' – the drawing together and gradual adhesion of matter thanks to gravitational forces.

ADAPTIVE GRADUALISM
A synthesis of evolutionary and geological theories, combining elements of Darwinian gradualistic thought and the concept that rapid evolutionary and environmental changes might occur on Earth during catastrophic episodes. Adaptive gradualism could be said to reflect the slow pace of evolutionary change interspersed with periods of rapid adaption following extreme environmental upheaval.

AERODYNAMIC FORCE
Otherwise referred to as aerodynamic drag and known to Newton as air resistance. The movement of an object such as a meteoroid through an atmosphere produces resistance, effectively slowing the object – a phenomenon known as drag. Drag is composed of three effects: skin friction due to rapid air flow over an object; base drag, generated by air disturbances behind the object; and finally, bow resistance, which sees air pressure waves building up in front of the object.

AIRFORCE SPACE COMMAND, USA (ASC)
Peterson Air Force Base, Colorado, operates strategic missile units and warning systems. The ASC reports to the Joint Chiefs of Staff in the US Department of Defense.

ALBEDO
Measurement of an object's reflectivity. Cloud cover, dust particles and vegetation all contribute to the albedo of a planet. Whereas

Earth reflects 35 per cent of all light received from the sun, Mars has an albedo of only 15 per cent because of its lack of vegetation and oceans, and its thin atmosphere. Albedos are a good measurement of an asteroid's size, assuming certain compositional characteristics are known. It is invariably measured in the infrared spectrum.

ANGLE OF INCURSION
The angle at which a would-be impactor enters Earth's atmosphere is critical. Doctor Schultz of Brown University, Rhode Island, believes that the dinosaur-killing K/T impactor had an oblique angle, thus throwing more energy into the atmosphere than into the ground and causing more widespread damage (especially across the northern hemisphere).

ANGLO-SAXON CHRONICLE
Maintained initially by monks and later endorsed and embellished by King Alfred the Great (849–99), it is the foremost historical document on Anglo-Saxon Britain and was maintained for three centuries, until 1154.

ANTHROPOLOGY
The scientific study of humankind, our origins and physical, social and cultural characteristics.

ANTIBALLISTIC MISSILE TREATY, 1972
As part of the Strategic Arms Limitations Talks of 1972, the Soviet Union and the USA agreed to cap the number of antiballistic-missile (a ballistic missile designed to destroy another ballistic missile in flight) sites to one hundred and to review this every five years. The treaty may well have been breached by Ronald Reagan's Strategic Defense Initiative (SDI) programme and can be withdrawn from with six months' notice.

ARIANE FIVE
$9 billion (US) launch vehicle developed by the European Space Agency (ESA) in association with the French space agency and members of the European Union. The vehicle was designed to be highly reliable, powerful and capable of carrying astronauts. A life-support capsule is still under development.

238

ARTHROPOD
Invertebrates of the phylum Arthropoda family, including crustaceans, insects, arachnids and centipedes.

ASTEROID
Often called small planetoids, they can be described as any small celestial body that orbits about the sun. These objects are typically between 1 and 670 kilometres in size. Composed of iron and rock, they are as old as the solar system itself: approximately 4.5 billion years old.

ASTEROID BELT
A great swath of small bodies forming a belt 2.1 to 3.3 AU from the sun, between the orbits of Mars and the gas giant Jupiter.

ASTRONOMICAL UNIT (AU)
149.6 million kilometres is equivalent to one Astronomical Unit, the average distance of the Earth from the sun. The AU is used as the standard measure of distance in astronomy.

ASTRONOMY
The scientific study of planets, stars, galaxies, interstellar and intergalactic space and the universe as a whole.

ATMOSPHERE
A gas and particle envelope surrounding Earth and many of the known planets. Earth's atmosphere sustains life, its most notable molecules being oxygen and carbon dioxide.

ATOM BOMB (A-BOMB)
A type of bomb in which the energy is provided by nuclear fission: atomic nuclei are split, releasing vast amounts of energy and radioactive particles. The first A-bombs were developed by the Allies in World War II using diverse research from scientists around the globe, including that of Otto Hahn (1879–1968), Albert Einstein (1879–1955) and JR Oppenheimer (1904–67). The Los Alamos, New Mexico, project, which developed the first atomic bomb, grew out of initial research carried out by Imperial Chemical Industries (ICI) for the British government.

AZTEC EMPIRE
An elaborate empire which ruled over 5–6 million people in the fifteenth and sixteenth centuries, based around the ancient Mexican capital city of Tenochtitlán. It was overthrown in the sixteenth century by the Spanish conquistadors. This society of Indian-Mexicans had an impressive organisational structure based around pagan beliefs and led by an appointed emperor. Intricate irrigation schemes were developed to maximise the available fertile land for agricultural production. Their cities were elaborate and highly decorative, often distinguished by the use of step-pyramid temples.

BALLISTIC MISSILE EARLY WARNING SYSTEM (BMEWS)
Operational since circa 1962 (the period of the Cuban Missile Crisis), the system depends upon Very Large Radar arrays to detect hostile missiles that might penetrate the USA.

BETA-TAURID STREAM
A complex of asteroidal material in an orbit that intersects Earth's orbit biannually. It is thought to be created by the emission of debris from Comet Encke.

BILLION
One thousand million: 10^9.

BIOGENIC
Materials derived from a living source.

BIOSPHERE
An area capable of sustaining life. On Earth it is a narrow band extending only a few metres below ground and approximately ten kilometres above it.

BOVINE SPONGIFORM ENCEPHALITIS (BSE)
A disease that destroys the brain tissue of cattle (its human form is New Variant Creutzfeldt-Jakob disease). Areas of the brain become spongy and nonfunctional, resulting in palsy, madness and death. Cases were first reported in Britain following the use of sheep protein to feed ordinarily herbivorous cattle, in the period between 1978 and 1988. It is believed that BSE is passed to humans, crossing the species barrier, via the consumption of infected beef.

240

BRITISH BROADCASTING CORPORATION (BBC)
A publicly funded broadcaster, formerly a private company, established under Royal Charter in 1927.

CASSIODORUS
A leading member of the Roman intelligentsia, Flavius Magnus Aurelius Cassiodorus (c.490–c.580) became a leading scribe and unwitting social historian.

CATASTROPHISM
Modern catastrophism is an outgrowth of a theory that fell from popularity in the eighteenth century. It blames catastrophic incidents for geological and evolutionary change.

CENTRAL INTELLIGENCE AGENCY, USA (CIA)
Set up to gather and sift intelligence about foreign powers, this federal US bureau was established by the National Security Act of 1947. One of the world's leading intelligence-gathering bodies, it provides information to the President and National Security Council of the USA.

CHARGE-COUPLED DEVICE (CCD)
A solid-state light/image sensor: a microchip designed to be used in video and other image-capture instruments. It is often used in optical astronomy. (See also NEAT.)

CLEMENTINE II
NASA's technology-proving space mission which is designed to intercept near-Earth asteroids and test surveillance, tracking and other systems.

CLIMATOLOGY
The study of a planet's climate, a science that originated in Ancient Greece and reached full maturity when German climatologist Wladimir Koppen (1846–1940) mapped the world's climatic zones and published his work in 1884.

COMETS
With a name based on the Ancient Greek word '*kometes*' meaning long-haired, comets are thought to be composed of ice and perhaps

loose agglomerates of stony matter and dust. They probably originate in a spherical (Oort) cloud of matter at the edge of our solar system. Comets are as old as the solar system itself. Long-period comets have orbital periods that last longer than 200 years; short-period comets (Halley-type) have orbits of less than 200 years.

COPERNICANISM
In 1543, Nicolaus Copernicus (1473–1543), a canon of the Polish Church, published *De revolutionibus orbium coelestium (On the Revolutions of the Heavenly Sphere)*, which rejected the theories of Ptolemy and Aristotle, who believed that the Earth was at the centre of all things. Copernicus placed the sun at the centre of our solar system with the Earth in orbit about it. This rejection of the notion that Earth was the centre of all things was nothing short of heresy. In 1616, Rome published an edict damning Copernicus's view, but others agreed with his theories (especially Galileo) and adopted Copernicanism as a working hypothesis.

CRETACEOUS PERIOD
144–65 million years BC: the last era of the dinosaurs, probably brought to a premature end by an impact with a vast comet/asteroid in the Yucatan.

CRO-MAGNON MAN
The first *Homo sapiens sapiens* (modern man), dating from 40,000 to 10,000 BC. They lived in rock caverns and makeshift shacks on the Russian and Central European Steppes. They practised deliberate, ornate burial of their dead and were highly artistic in nature.

DARWINIAN THEORY
Charles Darwin (1809–82) and Alfred Wallace (1823–1913) proposed that evolution is the result of natural selection of the fittest offspring, which is dependent upon variations in each individual. This variation or evolutionary advantage would be propagated as the fittest usurped the unfit.

DEEP SPACE 1 (DS1)
The experimental NASA spacecraft designed to test new ion-drive propulsion technologies. The craft may prove the value of new, long-

lived propulsion systems, thereby opening up the possibility for longer-term missions and NEO intercepts and trajectory changes.

DEFENCE EVALUATION AND RESEARCH AGENCY, UK (DERA)
Established in 1995, the agency forms the core of the UK Ministry of Defence's non-nuclear defence research assets. Guidance systems, electronic surveillance techniques, quantum optics incorporating tele-portation and cryptography, and countless other technologies have been devised and evaluated at DERA's facilities across the UK.

DENDROCHRONOLOGY
The use of annual tree-growth rates, demonstrated by concentric rings in tree-trunk cross-sections, to construct environmental and archaeological histories of a geographic area.

DEPARTMENT OF TRADE AND INDUSTRY, UK (DTI)
This department is geared to develop trade links abroad and supervise industrial diversity, while initiating training programmes and dissemi-nating knowledge of research carried out in the UK.

DESERT FOX
A military mission mounted by the USA and the UK in December 1998 against Iraq, in order to enforce United Nations resolutions.

DESERT STORM
A military operation initiated in January 1991 by a UN coalition in response to Iraq's invasion of Kuwait. The initiative ended in Feb-ruary 1991, having successfully repulsed Iraq from Kuwaiti soil.

DEVONIAN PERIOD
Period extending from 408 to 360 million years BC in which the first insects, molluscs and armoured fish evolved.

DNA
Deoxyribonucleic acid is the chain of encoded genetic information that allows cells to reproduce accurately in all organisms (except some viruses).

D-NOTICE, UK
This form of direct government dictat is generally issued in secret. However, in 1987 a BBC documentary film that revealed information about the existence of a secret British spy satellite known as Project Zircon was banned by Margaret Thatcher's government. The security services raided the office of the producer, Duncan Campbell, and forcibly removed virtually all material relating to the satellite. A High Court injunction forbidding Campbell to talk or write about the project was later repealed. As a result, in November 1987 selected audiences in both the UK and USA were allowed to view excerpts from the documentary. D-notices are generally more effective than in the case of the Zircon affair. Rarely used in modern times, they are more common in times of war or states of national emergency.

EARTH-CROSSING ASTEROID/OBJECT
Small bodies with orbits that approach the Earth or moon as a result of the gravitational influences of other planets.

ECOLOGY
A field of science that considers the symbiotic relationship of animals and plants within an environment.

ECONOMICS
A quantitative and qualitative scientific analysis of industrial, financial and social activity.

ECO-WARRIOR
An environmentalist predisposed to the use of radical and often dangerous tactics to protect an endangered environment.

EL NIÑO
A warm ocean current properly referred to as the El Niño Southern Oscillation. Its effects can distort seasonal weather patterns, resulting in drought during (what are ordinarily) rainy seasons, and in cyclonic storms, particularly between January and March. El Niño is known to occur every four to five years, but Global Warming may be generating a greater incidence.

EROS (438)
Eros is an asteroid targeted by NASA and Johns Hopkins University's Near Earth Asteroid Rendezvous (NEAR) mission.

EUROPEAN SPACE AGENCY (ESA)
The architect of Europe's exploration of space. It is funded by member states and its aims are purely scientific and commercial, with no military involvement or finance. ESA was formed in 1975 and its thirteen members include Austria, Belgium, France, Germany, Ireland, Italy and the UK. Its space base for rocket launches is located in Kourou, French Guiana.

EXTREMELY LOW FREQUENCY RADIATION (ELF)
Used by the US military to penetrate the ground and seawater for remote sensing and to transmit messages to submarines while submerged. Research by the High-frequency Active Auroral Research Program (HAARP) suggests that ELF waves could be employed as an antiballistic-missile technology in the near future.

GALILEO
Galileo Galilei (1564–1642) was an Italian astronomer, mathematician and physicist whose achievements included the perfection of the refracting telescope, which led to his discovery of Jupiter's satellites, sunspots, and craters on the moon. His advocacy of Copernicanism led to his being found guilty of heresy by the Inquisition in Rome. In June 1633 he was condemned to life imprisonment, which was later commuted to house arrest.

GEOSCIENCE
Any science, such as geology, geochemistry or geophysics, concerned with the Earth.

GIOTTO
The spacecraft that intercepted Halley's Comet in February 1986. Operated by ESA, it came within 605 kilometres of the nucleus, providing us with the first detailed view of a comet.

GLACIATION
The process by which a planet is covered with glaciers or masses of ice and annual snowfall exceeds annual melting and drainage.

GLOBAL POSITIONING SYSTEM (GPS)
Developed by the US military and completed in 1993, GPS depends on a network of 24 orbital satellites and is now employed as a navigation aid by travellers the world over.

GRAVITY MAPS
Built from the information provided by a gravimeter (an instrument for measuring the Earth's gravitational field at points on its surface), gravity maps show the variations in the gravitational field often caused by the varying densities of rock structure. This information is then used by geologists to build comprehensive maps of the Earth's surface, particularly in pursuit of oil reserves.

GREENLAND ICE SHEET PROJECT 2 (GISP2)
The second Greenland Ice Sheet Project was a National Science Foundation (NSF) project initiated in the USA. It ran from 1993 to 1998 and was concerned with the drilling and analysis of ice cores from Greenland's ancient ice sheets.

GREENPEACE
Founded in Canada in 1969, Greenpeace is the foremost environmental pressure group in the world, operating a number of strategies to preserve environments. One such strategy is the use of flotillas of ships as physical barriers to block the path of whaling vessels. The Greenpeace flagship is known as the *Rainbow Warrior*, of which there have been two to date.

GROUND BASED DEEP SURVEILLANCE SYSTEM, USA (GEODSS)
This system is designed to detect and monitor satellites and intercontinental ballistic missiles.

HALE-BOPP, COMET
A comet first discovered by modern man on 23 July 1995. A long-period comet, Hale-Bopp last visited Earth in circa 2214 BC.

HALLEY'S COMET
A comet that revolves round the sun in a period of approximately 76 years. This comet was named in honour of the English astronomer and mathematician Sir Edmond Halley (1656–1742), who correctly

246

predicted the 76-year orbital period of this remarkably beautiful comet.

HEAVEN'S GATE CULT
A quasi-religious sect, forty members of which committed mass suicide in March 1997, believing Comet Hale-Bopp to be the herald of their salvation. They sought a higher state of being and 'shrugged off' their bodies in order to ascend to another domain/stream of consciousness – or so they believed.

HIGH-FREQUENCY ACTIVE AURORAL RESEARCH PROGRAM (HAARP)
A US military research establishment based in Alaska. Officially designated as a research facility, it is apparently designed to gain a greater understanding of the ionosphere. However, patents allegedly filed relating to the project suggest that HAARP is more likely used for ionospheric manipulation. Using the HAARP, the remote sensing of enemy missile sites may well be feasible following the planned operational upgrade in 2002.

HOLOCENE
The second and current epoch of the Quaternary period (the most recent period of geological time, which began two million years ago and followed the Tertiary period). Holocene began 10,000 years ago at the end of the first epoch of the Quaternary period, the Pleistocene.

HOMO ERRECTUS, HOMO SAPIENS NEARDERTHALENSIS, HOMO SAPIENS SAPIENS
Homo errectus evolved approximately 1.8 million years ago on the African, Asian and European continents. His environment was diverse and challenging, undoubtedly requiring greater and better tool use for survival. Advanced tool use and increased dexterity, with improved inventiveness, led to the development of *Homo sapiens neanderthalensis* (180,000–35,000 BC). A highly skilled and evolved group, they appear to have been advanced and civilised, as the discovery of carefully buried disabled Neanderthals would suggest. *Homo sapiens sapiens* (modern man) developed contemporaneously with Neanderthal man, eventually superseding them and possibly even causing their extinction.

HURRICANES MITCH AND GEORGES
Hurricane Mitch was the fourth strongest hurricane recorded to date, achieving wind speeds of 180 miles per hour in September 1998. Hurricane Georges caused terrible devastation in the summer of 1998, particularly in Haiti, where 50 per cent of the country's food production areas was destroyed.

IMPACTOR
Any body, cometary/asteroidal/meteoroidal or man-made object that impacts with Earth or another planetary body.

INSECTA
Insects are part of the arthropod group, in the class of *Insecta*. Insects are readily distinguished from other creatures by the compartment-alised head, thorax and abdomen. They are by far the most populous creatures on Earth.

INTERNATIONAL SPACE STATION (ISS)
The ISS, also known as Alpha, will be a city in space. Construction in orbit began in 1998 and the station will inevitably be manned on a permanent basis. The project marks a turning point in international relations; it is so ambitious a concept, costing approximately $96 billion, that it can only be completed if there is genuine cooperation among the leading nations of the world.

IONOSPHERE
A zone of the Earth's upper atmosphere approximately 50–500 kilometres above sea level. This region contains a high concentration of free electrons, formed when the solar wind splits atmospheric molecules into electrons and positive ions, thereby 'ionising' them. This sphere of ions round the Earth protects the biosphere from the harmful effects of excessive ultraviolet radiation and X-rays.

K/T BOUNDARY
The interface between the Cretaceous period of geological time (144–65 million years ago) and the Cenezoec (65 million years BC to the present day). It is likely that the mass extinctions that occurred at this boundary were caused by a vast, catastrophic impact in the Yucatan. Following it, dinosaurs no longer walked the Earth, giving mammals a chance to thrive and dominate.

KUIPER BELT
It is possible that as many as 1 billion comets may exist in this orbital belt in the plane of the ecliptic beyond the orbit of Pluto, 35–1,000 AU from the sun. Comets within the Kuiper Belt may be perturbed by collisions with other bodies or by the gravitational influences of the gas giants, pulling them into the inner solar system.

LEONID SHOWER/STORM
An annual meteoroid shower that occurs in November (so named because the meteoroids appear to radiate from the constellation of Leo). At 33-year intervals, the Leonids are known to reach storm levels: perhaps 1,500 or more falling over Earth per hour. It is likely that 1999 will bear witness to one of these storms. Such a heavy rate of penetration may cause severe damage to orbital satellites, including the new International Space Station. The first great Leonid storm recorded in the modern age occurred on 12 November 1833, when hundreds of thousands of meteoroids were seen to fall over the USA.

LOS ALAMOS LABORATORY
The world's first nuclear weapons (atomic bombs) were constructed at Los Alamos, New Mexico, during the Manhattan Project of the early–mid-1940s. Two types of bombs were created: one based on a form of uranium ore concentrated to form a mass of the isotope known as uranium 235 (U235), which comprised less than 1 per cent of the natural uranium ore. The second bomb depended on plutonium, a synthetic material created in a reactor by bombarding uranium ore with neutrons. The U235 form of atomic bomb destroyed the city of Hiroshima in Japan on 6 August 1945; while the port of Nagasaki (also in Japan) was destroyed using the plutonium A-bomb. These two detonations helped the USA secure a declaration of surrender from the Japanese, effectively ending World War Two.

MAMMALS
Any animal of the *Mammalia* group, including humans, which have a number of distinctive characteristics: warm blood, a covering of hair, mammary glands in the female, red blood cells devoid of a nucleus, seven vertebrae in the neck, a four-chambered heart, and young that are born alive and nourished by their mother's milk.

METEOROID
A generic term for stony and/or metallic celestial bodies that are thought to orbit the sun, possibly as the remains of comets. Objects become known as 'meteors' when they blaze a fiery trail through our atmosphere, and are termed 'meteorites' when they hit Earth.

MILANKOVITCH OSCILLATIONS
Milutin Milankovitch (1879–1958) proposed that changes or oscillations in the Earth's orbit about the sun, from an ostensibly circular to an elliptical orbit, might lead to increases and decreases in solar radiation on Earth. Milankovitch suggested that this might be the cause of the warmer and cooler eras that have featured in geological history. It is possible that these oscillations can be used to explain the onset of ice ages.

MINISTRY OF DEFENCE, UK (MOD)
Britain's Ministry of Defence was formed in 1964 when the War Office, the Admiralty and the Air Ministry were subsumed to form one large organisation. The MOD is responsible for all matters relating to the defence of the realm.

MINOR PLANET CENTER, US (MPC)
The MPC is part of the Harvard-Smithsonian Centre for Astrophysics, Massachusetts, in the USA. It performs the task of cataloguing and verifying the orbits and nature of small bodies – such as asteroids – in space. The MPC also plays a major educational role, disseminating information about astronomy and inherent dangers from near-Earth objects.

NATIONAL AERONAUTICS AND SPACE ADMINISTRATION, US (NASA)
North America's space agency was first established in 1958, under the tutelage of former Nazi rocket supremo Wernher von Braun (1912–77). By embracing Nazi rocket research (post-war), the USA was soon able to develop a ballistic-missile programme of its own, culminating in the manned moonshot in 1969. NASA is the world leader in space-related activities and has contributed greatly to our knowledge of the Earth. It has also forced the pace of technological innovation in the fields of electronics, aeronautics and industrial design.

NATIONALISATION
Nationalisation is a government procedure whereby private companies, invariably monopolistic utilities such as electricity providers, are taken into public (government) ownership. (See also BBC.)

NATURAL ENVIRONMENTAL RESEARCH COUNCIL (NERC)
A quasi-governmental body established in the UK to fund research into natural disasters and environmental change. NERC administers grants and funding proposals for environmental research projects that will benefit the community as a whole. It could therefore play a part in funding research into the impact threat. However, NERC's current remit is deemed too narrow by its executives to allow it to fund work in this area.

NEAR EARTH ASTEROID PROSPECTOR (NEAP)
The NEAP spacecraft is the grand adventure proposed by SpaceDev, which is owned by US entrepreneur Jim Benson. This spacecraft will be funded by commercial investors and will target asteroids in an effort to gather commercially useful information and prove technologies as required by its commissioners.

NEAR EARTH ASTEROID TRACKING SYSTEM (NEAT)
This NASA programme was set up as a result of increasing public alarm in the USA. This alarm was in part fostered by the successful release of two motion pictures, *Armageddon* and *Deep Impact*, which focused on the dangers of a major asteroid impact with Earth. The NEAT programme is only a beginning, however. More observation is needed, especially in the southern hemisphere, where an active NEO search scheme does not currently exist.

NEAR-EARTH OBJECT (NEO)
NEOs, such as comets and asteroids, are bodies found to be travelling on trajectories that are likely to bring them into the vicinity of Earth. NEOs are the subject of an ongoing search by professional and amateur astronomers alike. Estimates vary as to the actual number of 1-kilometre (or larger) NEOs, although that figure is currently estimated to be in excess of 2,000.

NEOLITHIC PERIOD
The cultural period that lasted in SW Asia from about 9000 BC to 6000 BC, and in Europe from about 4000 BC to 2400 BC. This late Stone Age period is characterised by primitive crop growing and stock rearing, and the use of polished stone and flint tools and weapons.

NEWTONIAN
Of, or pertaining to, the science of Sir Isaac Newton (1642–1727). This word is generally employed when referring to Newton's concepts of mathematics and physics, his scientific technique being based on sound methodology and experimentation, as expounded in his work *Principia*.

NEWTON'S INVERSE SQUARE LAW
A common mathematical law which was employed by Newton to explain his notion of gravity. This law states that gravity (g) diminishes in an inversely proportional way the further away one travels from a source (d). It can therefore be written thus:
$$g=1/d \ (2).$$

NORTH AMERICAN AIR DEFENSE COMMAND (NORAD)
A satellite-tracking group operated by the US military. The group monitors the orbits of approximately 2–3,000 man-made satellites about the Earth and reports imminent collisions via the US Air Force's Satellite Control HQ. Potentially hazardous space debris is also monitored in an effort to protect valuable satellites and manned spacecraft from high-speed collision. NORAD was established in 1957 under a cooperative effort between Canada and the USA. Its original mission was to protect Canadian and US air space, but with the beginning of the modern space era in 1960 this changed to encompass space defence responsibilities. The Space Detection and Tracking System (SPADATS) is NORAD's key observational tool.

NORTH ATLANTIC TREATY ORGANIZATION (NATO)
Established by the North Atlantic Treaty in 1949, NATO is an international organisation comprising Britain, the USA, Canada and thirteen European countries. Its aim was to defend the West from communist aggression at the onset of the Cold War. NATO now has broader defensive goals in a world replete with diverse threats from international terrorists and well-armed dictators.

NUCLEUS
A core, central or fundamental part or thing round which others are grouped. The word 'nucleus' takes on a slightly different definition depending on which science it relates to. In this book, 'nucleus' takes on the astronomical definition: the head of a comet, consisting of small, solid particles of ice and frozen gases that vaporise on approaching the sun to form the coma and tail.

OCEANIC IMPACTS
This is when a body impacts with the ocean – such as in the simulation by Sandia National Laboratories. A vast tsunami is the likely result of a large (approximately 1-km) impactor in the sea. Coastal areas would be swamped and it is possible that the resulting ejecta would travel round the globe causing widespread damage.

ORBIT
In astronomical terms an orbit is the curved path taken by an object, such as a planet, satellite or comet, as it moves round another celestial body under the influence of a gravitational field. Orbits are usually elliptical in nature, but can also be circular or parabolic. The planets within our solar system travel round the sun in elliptical orbits.

OZONE LAYER
Also know as the ozonosphere, this zone forms part of the stratosphere round Earth and is composed of the gas known as ozone (O^3). The Ozone Layer is produced by the action of the sun on oxygen (O^2) in the air. Ozone absorbs harmful elements of the sun's radiated energy, protecting animal and plant life from excessive exposure to ultraviolet light. The Ozone Layer is degraded by chemicals such as chlorofluorocarbons (CFCs), but its generation is also inhibited if solar radiation is reduced. This would occur in an impact winter, when the atmosphere would fill with particulates, reflecting the sun's energy away from the Earth and changing its albedo.

PHANEROZOIC
The part of geological time represented by rocks in which the evidence of life is abundant. On the geological timescale this period comprises the Palaezoic, Mesozoic and Cenozoic eras.

PHOTOSYNTHESIS
The process by which plants and certain bacteria use light energy to convert water and carbon dioxide into simple carbohydrates, liberating oxygen into the atmosphere in the process.

PLASMA
In terms of physics, plasma is a hot ionised material consisting of nuclei and electrons, and is sometimes regarded as the fourth state of matter. It occurs on the Earth in gaseous forms (such as lightning) and in the upper atmosphere as part of the ionosphere. Much of the matter in interstellar space is believed to exist in a plasma state.

PLEISTOCENE
The first epoch of the Quaternary period, the Pleistocene lasted for about 1,990,000 years, extending from 2 million years BC to 8,000 years BC. It was characterised by extensive glaciations of the northern hemisphere and bore witness to the evolutionary development of man.

PLIOCENE
The last epoch of the Tertiary period, the Pliocene extended from 5.1 to 2 million years ago, immediately preceding the Pleistocene. During this period, many modern mammals appeared.

PROJECTILE
An object or body thrown in a specific direction by some form of energy transfer.

QUATRAIN
A stanza or poem containing four lines. The work of Nostradamus was written in this style.

ROSETTA MISSION
An ESA mission aimed at placing a robotic lander on the surface of Comet Wirtanen (pronounced 'Virtenen') in 2013. The lander, known as *Champollion*, was named in honour of Jean François Champollion, the man largely responsible for the translation of the Egyptian hieroglyphics using the Rosetta Stone.

ROYAL GREENWICH OBSERVATORY (RGO)
One of the oldest observational-astronomy organisations on Earth. First established by King Charles II in 1675, the RGO was formally disbanded in late 1998. This saw an end to more than 300 years of scientific endeavour and excellence at a time when more astronomy is desperately needed. Funding issues and shortsighted politicians were largely to blame.

SATURN V ROCKET
The 111-metre spacecraft that carried man to the moon, launched 16 July 1969. Fully laden, it weighed 2.9 million kilogrammes. Its five Rocket F-1 engines produced 3.45 million kilogrammes of thrust.

SEASONAL ADJUSTMENT DISORDER (SAD)
Also known as Seasonal Affective Disorder. A depressive state affecting some individuals during the autumn and winter months. Additional exposure to daylight can alleviate symptoms.

SCHMIDT TELESCOPES
Astronomical telescopes used to make distortion-free photographic surveys of the sky over a wide field. This form of telescope was invented by the Estonian Bernhard Schmidt (1879–1935), and is in common use to this day.

SEARCH FOR EXTRATERRESTRIAL INTELLIGENCE (SETI)
NASA established a SETI programme in 1992, but the US Congress withdrew funding only one year later. Work now continues in the private sector. Data gathered via radio telescopes is being decoded by scientists around the world. An innovative initiative whereby members of the public can download sections of data and specially designed processing software has also encouraged people to use their home PCs to process radio emissions from the stars in the search for extraterrestrial signals. When the processing is complete, they send the data back to SETI – in essence vastly increasing SETI's processing power without the need to seek specific funding for additional computers of its own.

SHIELD
A research project mounted in late 1998 by Doctor Robert E Gold of Johns Hopkins University. Its aim was to devise the most likely

NEO hunter-killer system. Shield is envisaged as a multi-tiered robot detection and destruction/redirection system that would operate in space to prevent a meteoroidal/asteroidal/cometary collision with Earth.

SILENT SPRING
Rachael Carson published this seminal work in 1962. The book espoused the dangers of pesticide overuse and had the effect of a rallying call for like-minded individuals, fostering the development of the modern environmental movement.

SOLAR SYSTEM
A system containing a sun and other bodies (such as planets, comets and other materials) held in a gravitational field. Our solar system is composed of nine planets: Mercury, Venus, Earth, Mars, Jupiter, Saturn, Uranus, Neptune and Pluto. The sun accounts for 99.86 per cent of the mass of the solar system. Jupiter composes 66 per cent of the remaining mass, hence its significant effect on the orbits of small bodies.

SOLAR WIND
A stream of charged particles emitted by the sun at high velocities. Its intensity increases during periods of heightened solar activity. It affects terrestrial magnetism, some of the particles being trapped by the magnetic lines of force, and causes auroral displays.

SPACEGUARD
A pressure group established by like-minded individuals following the impact of Comet Shoemaker-Levy 9 with Jupiter in 1994. Spaceguard's primary objectives are the cataloguing of all potentially hazardous objects in space and the education of the public and governments alike as to the impact threat.

SPACETRACK
A planet-wide network of cameras and radars, operated by the US air force, specifically developed to track missiles and satellites. The system was the first to use phased array radar, based initially at Eglina Airforce Base, Florida. Initiated in 1963, the programme became partially operational in 1969.

STRATEGIC DEFENSE INITIATIVE/STAR WARS (SDI)

In the 1980s, Ronald Reagan's administration of the US government proposed a programme of space-based detection and deflection technologies. Known as SDI, and later nicknamed Star Wars, this initiative outlined the implementation of a defensive shield to protect the USA from foreign intercontinental ballistic missiles fitted with nuclear warheads. The SDI is a secret project by nature and much of the detail regarding the outcome of the costly research (estimated by some sources to exceed $130 billion) is obscured. Its development and deployment would certainly contravene numerous nonproliferation and outer-space treaties, and as a result the specifics of SDI and its deployment are likely to remain shrouded in secrecy. However, it is well known that some of the software designed for missile interception was given its first outing during Operation Desert Storm in 1991. It is highly likely that SDI development work continues to this day, despite the Clinton administration's dislike of the initiative. The air force and Pentagon's ballistic-missile defence group announced on 9 February 1998 that they had awarded the first portion of a $3 billion (US) contract to fund research into the possibility of space-based lasers being used to destroy ballistic missiles launched by enemy states.

STRATOSPHERE

The layer of the Earth's atmosphere between 15 and 50 kilometres above the Earth that is stable and unaffected by weather. Contains what is commonly known as the Ozone Layer. (See also Ozone Layer.)

SUBMARINE LANDSLIDES

Tsunamis are often triggered by submarine landslides. The flooding of Papua New Guinea on 17 July 1998 was probably the worst of the twentieth century and was most likely triggered in this way. The tsunami killed at least 3,000 people either by drowning, or as a result of the mudslides and land collapse that followed the wave.

SUPERHEATED EJECTA

The debris thrown away from an impact site by tremendous energy exchange. Air friction heats the rock, water and other material to very high temperatures. For example, 70 seconds after the Jupiter collision, ejecta debris reached temperatures of approximately 2,000

257

degrees Celsius, 100 times hotter than Britain's average July temperature (17–20 degrees Celsius).

SUPERNOVA
A star that explodes due to instabilities caused by the exhaustion of its nuclear fuel. For a few days after the explosion, the star will be up to 100 million times brighter than our sun. The expanding shell of debris emits radio waves, X-rays and visible light for hundreds or thousands of years.

TECTONIC EVENTS
'Tectonics' is the study of the process by which the Earth's surface attained its present structure. Specific tectonic events can be identified as being responsible for forming distinct structures, providing a 'cause and effect' equation.

TERAFLOPS
Teraflop supercomputers are used by the Sandia National Laboratory to perform Shock Physics computations (simulating nuclear detonations). Sandia have employed the Teraflops to simulate the effects of comet impacts on Earth. Under a planned programme of updates, the Teraflop computers are expected to have a 9,000-processor configuration in the near future.

TERRESTRIAL
A term used to describe something as being of, or relating to, the Earth.

TERTIARY PERIOD
The first period of the Cenozoic era, which lasted for 69 million years, a period that saw mammals rise to dominance.

TNT
TNT stands for 2,4,6-trinitrotoluene, a yellow solid used chiefly as a high explosive. It is used as a standard measure for all explosive potentials (i.e. a hydrogen bomb might have the explosive ability of 10 megatons of TNT). It is also an intermediate in the manufacture of dyestuffs.

TRAJECTORY
The path described by an object moving in air or space under the influence of one or more of the following forces: thrust, wind resistance and gravity. Projectiles commonly follow a curved trajectory. (See also Projectile.)

TSUNAMI
Sea waves produced by undersea disturbances, such as earthquakes, subsidence, volcanic activity, or by impactors hitting the ocean. These waves are often extremely destructive, engulfing coastal areas and cities. Humankind has little or no defence against such waves. (See also Submarine landslides.)

ULYSSES MISSION
A mission to observe the sun and solar winds from a high-solar-latitude perspective. It requires the spacecraft to fly on a trajectory that passes over the poles of the sun. Originally named the International Solar Polar mission, this is a joint venture between NASA and ESA, and was launched in 1990.

Index

Aasaro, Frank 15
Adams, J 53
1967 Agreement Governing the
  Activities of States on the Moon
  and Other Celestial Bodies 91
Alaska 100
Alaskan gas fields 103–4
albedo 36, 58
Alpha space station 72–3
Alvarez, Luis 15, 16
Alvarez, Walter 15–16
Antarctic ice sheet 58–9
Antarctica 60
1972 Antiballistic Missile Treaty 91
Applewhite, Marshall 75, 79–80
Arco 103–4
Arrhenius, Svante August 30
Asher, DJ 35, 37
asteroids xxviii–xxix, 21, 114
  Ceres xxix
  Earth crossing 3
  Eltanin 60
  433 Eros 126, 127
  1996 JA1 3, 95, 108–9
  1992 KD 122
  4660 Nereus 138, 139
  1988 OH 106
  1988 OR2 106
  (4179) Toutatis xviii–xix
  1997 XF11 xiv–xviii, 142–4,
    145, 147
Astronomical Unit xxviii–xxix

Atlantis legend 44–5, 48
atomic bombs 93
aurora borealis 100, 104

bacteria 29–30
Baillie, Dr Michael 9, 39, 47,
  61–2, 64
Ballistic Missile Early Warning
  System (BMEWS) 97, 99
Barnish, SJ 40
Barringer crater 9
Barringer, Daniel 9
Battle of Hastings 74
Battle of Stamford Bridge 74
Belleek explosion 70–1, 86
Bellingshausen Sea 60
Benfield Greig Hazard Research
  Centre 37
Benson, Jim 137–9, 140, 141–2,
  147
HMS *Bermuda* 93
Beta-Taurid Complex 22, 35, 38, 50
biogenic comets *see* under comets
Bogotá fireball 86
Bopp, Thomas 75
Borreally comet *see* under comets
Boslough, Mark 25
Bottacione Gorge 15
Brazilian forest impact 80–1
Breckenridge, Arthurine 24
BSE 8, 85–6, 92

carbon dioxide 14, 37–8, 54, 62
Carnegie model 54
Carson, Rachel 89, 146
Casanova, Bob 129
Cassiodorus, Magnus Aurelius 39–40, 47
catastrophism xxvi, xxxi
Ceres see under asteroids
Champollion 116, 117
Channel Tunnel 154
Chiba model 31–2
Chiba, Satoshi 31–2
Chicxulub crater see under craters
Christmas Island 93
CIA 98–9
Clementine II 153
Clube, Victor 15, 35, 37, 38, 50
comets xxx–xxxi, 3, 29, 114
  biogenic 29
  Borreally 122
  Encke 22, 35, 38
  Hale-Bopp 75, 79, 114
  Halley's xxxi, 30–1, 75, 78, 95, 114, 115–16, 120
  proto-Encke 35–6, 37
  Wild-2 31
  Wirtanen 114, 116, 117, 121
Conquistadors 73
Copernicus, Nicolaus xxiii
Cortés, Hernando 73
craters xiii, 82
  Barringer 9
  Chicxulub xiii, 17–18, 19
Crawford, David 24, 25
Cretaceous/Tertiary boundary see K/T impact
Creutzfeldt-Jakob disease (CJD) 8, 9, 92
Cuban Missile Crisis 103
Cudaback, David 15
Cuvier, Georges xxvi

Debbenham, Scott 58

Deep Space 1 122–3
Defence Evaluation and Research Agency 110
dendrochronology 19–20, 38–9, 40, 61–2
Desert Fox 96, 108, 121
detection/defence systems 23, 73, 94–108, 113–33
  SDI 4, 90–2, 96–7, 103, 121–2
dinosaurs xiii, 12–16, 48
Drake, Edwin 141
dust rings 53–4

earth tremors 37
Earth-crossing asteroids (ECAs) 3
earthquakes 7, 63
economy 25–6, 67–9
El Niño 33, 52, 154
electromagnetic pulse 102, 103, 104
Eltanin see under asteroids
Encke see under comets
Encke, Johann Franz 22
energy release 7
environmental stability 32
433 Eros see under asteroids
Eskimos 46, 55
European Space Agency (ESA) 30, 126, 140, 152, 153
European Union 83–5
Extremely low frequency (ELF) waves 100

Feduccia, Alan 12
ferns 18–19
fire, as result of impact 7, 14, 52, 86
fireballs see meteoroids/meteorites/ meteors
floods 57, 58–9, 61–4, 65

Gagarin, Yuri 72
Galileo Galilei xxiv
Galileo probe 155
GEODSS see Ground-Based

Electro-optical Deep-Space
   Surveillance System
giant ground sloths 56
Giotto mission 30–1, 115–16
GISP2 *see* Greenland Ice Sheet
   Project II
global cooling 14, 18, 35, 36,
   51–3, 54
global warming 14, 36, 37–8, 54–5,
   62, 65
Gold, Dr Robert E 129–31
Goldin, Dan 126
governments
   United Kingdom 8–9, 92
      gas attacks 95
      impact threat 107–8
      letter to John Major 94
      NERC 109–10
   United States 5–6, 96–9
      Clementine II 153
      disaster announcements 95
      HAARP 100–1, 104–5
      ionosphere 102, 103
      NEAR 126–8
      NEAT 105–6
      Pacific explosion 107
      SDI 4, 90–2, 96–7, 103, 121–2
      Shield 129–31
gradualism xxvi, xxxi
Grady, Dr Monica M 6, 7–9,
   17–18
gravity maps 5
Great Depression 26
greenhouse effect 54, 62–3
Greenland 46, 87, 99
Greenland Ice Sheet Project II
   (GISP2) 20–1, 22, 46
Ground-Based Electro-optical
   Deep-Space Surveillance System
   (GEODSS) 97, 105
Gulf War 92
Gustafson, Miriam 147

HAARP *see* High-Frequency Active
   Auroral Research Program
Hadrata, Harald 74
Hale, Alan 75
Hale-Bopp Comet *see* under comets
Halley's comet *see* under comets
Halley, Sir Edmond 75
Halophiles 29–30
Hands-On Universe Program
   147–8, 149
Harold II (Harold Godwinson)
   74–5
Heaven's Gate cult 75–6, 79–80
Heinrich events 53
Heinsohn, Gunnar 47
Hekla 4 volcanic eruption 62
Helin, Eleanor 142, 149
Hergenrother, Carl 3
High-Frequency Active Auroral
   Research Program (HAARP)
   100–1, 104–5
Hildebrand, Alan 17
Hiroshima 93
Hogg, Douglas 92
Holdridge, Mark 126–8
Homer 43
Homo sapiens
   neanderthalensis 32–4, 36, 63–4
   sapiens xxii, xxv, xxvii–xxviii,
      xxxii, 32–4, 36
Hoyle, Sir Fred 28–30
Hudson River 47
Human Genome Project 64
Hunter B 111
hurricanes 37
   Georges 111
   Mitch 64, 67, 68, 111
Hutton, James xxvi
Hybrid Rocket Division 139
hydrogen bomb 4
hyperinflation 26

ice ages 35–8, 46, 49–51, 54, 55,
  57–9, 65–9
ice-core records 20–1
impact winter 6–7, 65
Indonesian fires 52
Industrial Revolution 38
Inquisition xxiii–xxiv
1967 International Concord 91
International Development
  Corporation 111
International Earthshield Treaty
  132
International Space Station 125,
  126
Internet 146–7, 149
Inuits 22, 46, 55, 76–7, 79
ion drives 122–3
ion thrusters 129, 130
ionosphere 100, 101–4
ionospheric heaters 103, 104
iridium
  fingerprints 20
  layer 16, 18
  Tunguska blast 22

1996 JA1 see under asteroids
Joergensen, Bjoern Franck 87
Jupiter
  collision 25, 27, 82–3
  Galileo probe 155

1992 KD see under asteroids
Kepler, Johannes xxiv
K/T impact 10–21, 32, 47
Kuiper Belt xxxi, 147, 148, 149

Leonids 151–2
Levy, David 27, 82
Limited Arms Test-Ban Treaty 103
Little Ice Age 46
Lockheed Martin 23–4
Lovell, Sir Brian 102
Lucretius 152

Lumley, Joanna 135
Lunar-Sat 140–1, 142
Lyell, Sir Charles xxvi

magnetosphere 100, 101
Maiasaura 12
Major, John 4, 94
Manhattan Project 4
Mantell, Gideon Algernon 48
Mars mission 129
Marsden, Brian 142, 144, 145,
  149–50
Maslin, M 53
Mason, Tom 70–1
mass extinctions 10, 12–16, 32
master-chronology 20
Maui telescopes 105
McCurdy, Heather 147
Menwith Hill Station 97–9
meteoroids/meteorites/meteors
  xxix–xxx, 3, 86
migrations 46–7
Milankovitch, Milutin 50–1
Milankovitch oscillations 53–4
Miller, Rich 15
Ministry of Defence (MOD) 8, 9,
  94, 110
Montezuma II 73
Mount Pinatubo 51–2
Muinonen, Karri xvii, xviii
Mycenaean civilisation 42–5
myths 41–5, 47–8, 61–3

Nagasaki 93
Napier, William 15
NASA xviii, 23–4, 27, 83, 131, 153
  Deep Space 1 122
  Mars mission 129
  NEAR spacecraft 126
  NEAT programme 105
  Stardust mission 31
National Science Foundation 20,
  147–8

Natural Environmental Research Council (NERC) 109–10
Neanderthal man *see under Homo sapiens*
Near-Earth Asteroid Prospector (NEAP) 138–40, 142
Near-Earth Asteroid Rendezvous (NEAR) spacecraft 126–8
Near-Earth Asteroid Tracking programme (NEAT) 99, 105–6, 112
4660 Nereus *see under asteroids*
Newton, Sir Isaac xxv
nitric acid 14
nitric oxides 6–7
Noah 61, 62, 64
Non-Proliferation Treaty 107
Norse migrations 46
North American Defense Command (NORAD) 97
North Korea 107
Northfield Mount Hermon School 147
Nostradamus 113
nuclear deflection system 119–20, 123–5
nuclear detonations 93
nuclear-pulsed rockets 123–4
Nuuk 87

1988 OH asteroid 106
oil rush 141
Olympus spacecraft 152
omens and fears 73–81, 83
Oort cloud xxxi
Oppenheimer, Robert 4
1988 OR2 asteroid 106
Ostrom, Professor John 12
1967 Outer Space Treaty 124
Out-of-Africa theory 33, 64
output failure 111
Ovid 41–2, 45, 61, 62, 63
ozone layer 52

Pack, Hughes 147
Panspermia 28–31
Paris nations 68
Patriot missiles 92
Pave Paws 97
Pennypacker, Carl 148
perpetual winter 51
Perseids 152
Peterson, George 147
Phaethon 42, 43, 44–5, 47
Phanerozoic period 59
phytoplankton 14
Pinatubo eruption 51–2
Pisco 60
Plato 43–4, 45, 48
Pleistocene Ice Age 49–50
FS *Polarstern* 60
popping-cork effect 59
potentially hazardous asteroids (PHAs) xiv, xvii, xviii–xix
Pravdo, Dr Steven 105–6
proto-Encke *see under comets*

race violence 34
rainfall 54–5
Rapid Reaction Committee 132
risk calculations 7–9, 87–8, 110
Rojas, Captain Carlos Augusto 86
Rosetta mission 116–17, 121
Royal Greenwich Observatory 110
Rutherford Appleton Space Laboratory 115

Saddam Hussein 92, 96, 100, 121
Sandia National Laboratories 24–5
Saunders, Dr Mark 37–8, 54
Schliemann, Heinrich 45
Scotti, Jim 142
Scud missiles 92, 96
SDI *see* Strategic Defense Initiative
sea
    levels 49, 54, 60, 62–4
    salinity 53

Search for Extraterrestrial
  Intelligence (SETI) 149
seasonal affective disorder (SAD) 67
Shield 129–32, 153–4
shockwaves 7, 13, 22, 64
Shoemaker, Carol 27, 82
Shoemaker, Eugene 27–8, 82, 109
Shoemaker-Levy 9 25, 27, 82–3
siege mentality 69
Silent Spring 89, 146
simulations 24–5
Sitarski, Grzegorz xviii
slingshot manoeuvre 118
solar wind 101
Space Command 91, 97, 105, 107,
  108–9
Space Detection and Tracking
  System (SPADATS) 97
Space Guard 109, 150
Space Innovations Ltd 138
Space-Based Infrared System
  (SBIRS) 97–8, 99
SpaceDev Incorporated 137–40
Spahr, Tim 3
Spuck, Tim 147
Star Wars see Strategic Defense
  Initiative
Stardust mission 31
Stewart, Joseph 148
Storey, Brian 58
Stracken, Henry Joy 86
Strategic Defense Initiative (SDI) 4,
  90, 91–2, 96–7, 103, 121–2
student protests 151
sulphides 18
sulphur dioxide 14
Surrey Satellite Technologies Ltd
  (SSTL) 140–1

Tate, Jonathan 150
tektites 19

Teller, Edward 4–5, 24, 94
Temple-Tuttle 152
teraflops 24
Tesla, Nikola 101–2, 103–4
Thomas, E 53
Thule Eskimos 46
(4179) Toutatis see under asteroids
tree-ring data 20, 39, 64
tsunamis 7, 19, 60, 63
Tunguska impact xiii, 22–3, 76–8,
  79, 81–2
Turner, Ray F. 115–19, 125

Ulysses mission 118
uniformitarianism xxvi–xxvii
Urias, Colonel John M 5–6, 26
Ussher, Bishop xxvi

Van Allen belts 102, 104
Vermeersch, Piere 33
Vikings 46
violence 67–8
Vita-Finzi, Professor Claudio 55
volcanoes 20, 37, 41, 51–2, 59, 62

Wall, Dr Jasper 22–3, 87–8, 89,
  110, 111, 119, 132–3
Welles, Orson 95
Whipple, Fred L xxx
Wickramasinghe, Chandra 28–31
Wild-2 see under comets
William the Conqueror 74
Wirtanen see under comets
Woolley, Charles Leonard 62–4

xenophobia 36
1997 XF11 see under asteroids

Yeomans, Donald 142, 143
Yucatan Peninsula 13–14, 17

Zielinski, Gregory 59